Sex Robots & Vegan Meat

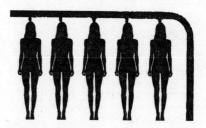

Sex Robots & Vegan Meat:

Adventures at the Frontier of Birth, Food, Sex & Death

JENNY KLEEMAN

PICADOR

First published 2020 by Picador
an imprint of Pan Macmillan
The Smithson, 6 Briset Street, London EC1M 5NR
Associated companies throughout the world
www.panmacmillan.com

ISBN 978-1-5098-9488-8 HB
ISBN 978-1-5098-9490-1 TPB

1 3 5 7 9 8 6 4 2

A CIP catalogue record for this book is available from the British Library.

Typeset in Warnock by Jouve (UK), Milton Keynes
Printed and bound by CPI Group (UK) Ltd, Croydon, CR0 4YY

Visit **www.picador.com** to read more about all our books
and to buy them. You will also find features, author interviews and
news of any author events, and you can sign up for e-newsletters
so that you're always first to hear about our new releases.

For Benjamin and Isabella, of course

'We're going to enter a very competitive market, a lot of players. We think we're going to have the best product in the world. And we're going to go for it and see if we can get 1% market share.'

Steve Jobs at the launch of the first iPhone, 9 January 2007

'Wanting to reform the world without discovering one's true self is like trying to cover the world with leather to avoid the pain of walking on stones and thorns. It is much simpler to wear shoes.'

Ramana Maharshi

Contents

Preface

What you are about to read is not science fiction.

We are on the brink of an age when technology will redefine birth, food, sex and death – the fundamental elements of our existence. Until now, human life has always meant emerging from our mother's body, living off the flesh of dead animals and seeking out sexual relationships with other humans, until it ends with a death we can neither avoid nor control.

Over the past five years, I've immersed myself in the world of four different inventions that promise to deliver the perfect partner, the perfect gestation, the perfect meat, the perfect death. They aren't perfect yet; they are still being made, in labs and garages and studios, in hospitals, workrooms and warehouses. Some of them will be on sale within a few years, others will take decades to be market ready, but they will all become an inevitable part of human life.

How much are we about to hand over to technology? And how will it change us? To answer this, we will travel across four continents and visit the darkest parts of the internet. I am going to take you to the kitchens where priceless chicken nuggets are made, the members-only meetings where people learn how to kill themselves, the labs where foetuses grow in bags and the discussion boards where men plan all-out war against women. We will meet scientists, humanoids, designers, ethicists, entrepreneurs and provocateurs; we will meet a fertility specialist prepared to do almost anything to satisfy his patients, a man who is married to his sex doll, a cake decorator who helped her best friend die and an artist who uses living flesh as his medium.

The men I'll encounter who are behind these technologies (and they will pretty much all be men) are sometimes driven by principle, sometimes by passion, often by money, but always by the promise of validation and fame. They all share the belief that technology can let us have the lives we truly desire without sacrifice; that it can eliminate our problems and set us free.

But even the smartest visionaries can't foresee where their innovations will take us. When Steve Jobs launched the iPhone, he dared to hope that it would capture 1 per cent of the market; he had no idea that smartphones would take over our lives, overshadow our relationships, become the external organ we can't function without. Radically disruptive technology always comes with aftershocks too extraordinary ever to predict.

If we can have babies without having to bear children, eat meat without killing animals, have the ideal sexual relationship without compromise, have a perfect death without suffering, how else will human nature be changed forever?

Without us realizing it, human existence is being redefined in ways no one can determine or control.

To show you why I believe this is already happening, let me take you to a Southern California factory, where the world's most glamorous adult toys are made.

PART ONE

THE FUTURE OF SEX

The rise of the sex robot

CHAPTER ONE

'Where the magic happens'

Abyss Creations is an unremarkable grey building off Route 78 in San Marcos, a thirty-minute drive north of San Diego, with a half-empty parking lot and a high perimeter wall. There's no sign, no logo, no indication that a world-famous, world-leading multimillion-dollar sex toy business operates behind the tinted glass. They don't want to attract passing customers, or fanboys, or rubberneckers.

When you pass between the sliding doors, you are greeted at reception by a seated life-size female doll in black glasses and a white shirt that strains to contain her heavy cleavage. A male doll stands beside her, dressed in a grey tie and waistcoat; his almond eyes and sharp cheekbones are unmistakably those I've seen in videos and photographs of Matt McMullen, Abyss Creations' founder, chief designer and CEO. There's a very convincing plastic orchid with sinuous fake roots creeping over onto the counter. Everything here is synthetic, but you only realize on second glance.

Abyss Creations is the home of RealDoll, the world's most famous hyperrealistic silicone sex doll. Every year, up to 600 of them are sent out from the workshop in San Marcos to bedrooms in Florida and Texas, Germany and the UK, China and Japan and beyond, costing anything from $5,999 for a basic model to tens of thousands if the customer has unusual specifications. *Vanity Fair* magazine calls them 'the Rolls Royce of sex dolls'. RealDolls have modelled in Dolce & Gabbana fashion shoots and starred in a string of movies and TV shows, from *CSI: New York* to *My Name is Earl*, and most famously opposite Ryan Gosling in *Lars and the Real Girl*. This is the most high-end masturbation on the market.

Dakotah Shore, Matt's nephew and all-round helper, is going to take me on a tour of the factory. He lopes up to shake my hand, a warm smile beaming out from behind a magnificent copper beard. Dakotah works in shipping and handles the social media accounts. He's only twenty-two, but he's been working here since he was seventeen. He's grown up alongside the dolls.

'My dad worked here when I was a kid. Matt's my mom's brother and I'm really close with him. So it's always been part of my life, it's never been weird to me,' he explains as he leads me behind reception, past a rack of dolls in lacy underwear and high heels. There is a blonde doll with porcelain-pale skin and glossy cherry-red lips, and a mixed-race doll with tumbling curls. A goth doll has studs in her nose, lip and navel, and bolts through her nipples clearly visible in her fishnet halter-neck. 'The first time I came here I was twelve or thirteen years old and I thought it was cool,' Dakotah continues, then checks himself. 'I didn't see the *whole* factory, I just saw the receptionist mannequins upstairs and I thought, Cool – really realistic receptionists.' He gives me a sheepish grin.

We walk along a corridor lined with framed press cuttings and movie posters featuring RealDolls. There's what looks like a Disney drawing, until you look closer and realize it's Snow White being groped by all seven dwarves. Dakotah props open a door with an enormous, veiny, erect silicone penis. 'Now that I work here and know the depth of it, it's normal to me. It's something that makes a lot of people happy and I take a lot of pride in it.'

We go down some steps that lead to the basement, passing underneath the giant labia of a huge doll that straddles the stairwell. She has bluish-grey skin and thick, articulated tentacles for hair; she was a prop in a minor Bruce Willis movie called *Surrogates*. At the bottom of the stairs is an enormous room with halogen strip lights. This is the beginning of the production line.

'This is where the magic happens.'

A long queue of headless bodies hang on metal chains from a track in the ceiling, like carcasses in an abattoir. Their fingers and legs are splayed, their chests jut forwards and their hips are thrust back. Each one is different to the next: some have cartoonish, pendular breasts, others have athletic bodies, but they all share the same impossibly tiny waists. Because they are hanging, they are moving, dangling eerily a few feet above a floor littered with gummy silicone offcuts that look like flakes of dead skin.

'You can touch them, no problem,' Dakotah says. He slaps one hard on the bottom. 'Sounds just like a human.'

It does. It makes me flinch.

The skin is the most unnerving thing of all about these headless bodies. Made of a custom blend of medical-grade platinum silicone, in a range of tones from *fair* to *cocoa*, it feels like human flesh, with the same friction and resistance, but it's cold. The hands have lines, folds, wrinkles, knuckles, veins. When I intertwine the fingers of one in my hand I can feel the crunch of the skeleton underneath, with joints, just like bones.

'Hands are the hardest thing to sculpt,' Dakotah tells me. 'We usually mould them off a real human's hands.' He stops to take a close look at a few. 'Actually, some of these hands belong to my ex-girlfriend.'

Mike is delicately snipping excess silicone off the seams of one of the hands with some tiny scissors. Brian is filling the moulds around the skeletons, ready for casting into the thrusting, ready-for-action poses. Tony is having a sandwich. There is nothing seedy about this workplace: it's a workroom, a factory, and for the technicians down here, the dolls are mundane. These guys might as well be assembling toasters.

Seventeen people work in the San Marcos HQ, but that is not enough to keep up with demand. From order to shipping, it can take more than three months to produce a RealDoll. The attention to detail, the skill involved in creating a doll, is undeniable. Dakotah is visibly proud of it all, and so earnest that I almost don't want to ask my next question. Because, even though these are RealDolls, there is little that's real about them. They have the bodies of surgically enhanced porn stars. They are caricatures.

'Women don't look like this, do they?' I say.

'We do have some that are 100 per cent modelled off real women, some are real, but yes, they are generally a little bit exaggerated,' Dakotah concedes. 'We like to make it the *ideal* female form.'

RealDolls are fully poseable, with a skeleton made of custom steel joints and PVC bones. They're designed so that the doll has a similar range of movement to a human – except for the legs.

'You can open them up pretty wide, and they go up pretty high,' Dakotah says, doing some high kick gymnastics with a headless doll, pulling an ankle up to a collar bone until I wince.

'A human can't do that,' I say.

'A real human can't do that, no. Well, some can, but not all.'

'But the perfect woman can do that?'

'The perfect woman could probably do that.'

The perfect woman has the waist-to-hip ratio of a Kardashian and the joints of a contortionist.

Dakotah takes me to a table covered with vaginal inserts, which are removable pink sleeves that go into the doll's vaginal cavity, like a kind of ribbed rubber sock with labia at the opening. 'We have fourteen different kinds of labia,' he says with a flourish. There are mouth inserts too, all with removable tongues and perfect teeth (bad teeth is one of the few things no one asks for, Dakotah says). The teeth are soft silicone, so there's no chance of anything pushed between them getting snagged.

In the early days, the only way to clean a used RealDoll was to take it into the shower or the bathtub. The invention of inserts was a game changer. 'You wash it in the sink. If you want it to be nice and soft, put some baby powder into it, but you don't need to. Then it just slides back in,' Dakotah tells me, as if he were describing how to change the bag on a vacuum cleaner. 'A lot of our customers have multiple inserts.'

There are male dolls, but not many. I spot one hanging on the production line, dressed in a surgical gown. His head is attached, and it's got the Matt McMullen doppelgänger face. He's looking down on us, with an expression that's supposed to be dark and brooding, but comes across as a bit snooty from a foot above my head.

'There's a male doll over there who looks a lot like Matt,' I say.

Dakotah looks up from the labia. 'It might be the Matt face. It's actually called the *Nick* face. He sculpted it himself, to be based on himself.'

'He sculpted his own face so that people can buy it and have sex with something that looks exactly like him?'

Dakotah hesitates. 'You can customize the face so it doesn't always look like him. It's just the structure of the face that looks like him.' For the first time since we met, he looks embarrassed.

He whips off the surgical gown – there to protect the doll from dust, because this one has been in the workroom for a while, he says – to reveal a very boyish, slim body, with a tight six pack, in white boxer shorts. He's far less realistic than the female dolls: instead of a wig, there's an approximation of stubble painted across his head, which makes him look like a very weedy Action Man. I have a feeling that these male dolls aren't designed for women at all. This model is young and skinny, what gay men might call a twink.

'Do women actually buy these?'

'Men and women buy them. More so men, but we do get female buyers,' Dakotah shrugs. 'For the dolls, I would say less than 5 per cent of the buyers are female. But we do sell accessories, all kinds of dildos, and many more women buy those. I think women are more likely to buy a toy than a full-on doll, for some reason.'

I have a hunch about what that reason might be. I try to imagine straddling one of these expensive, cold lumps of silicone. It would feel ridiculous, desperate, the opposite of erotic. Sex with anyone or anything that doesn't have genuine desire for me would not be sexy for me, and while I can't speak for all women, I don't imagine this is a minority view. A dildo isn't masquerading as a person, and you don't have to pretend it's really into you to enjoy using it.

'Maybe because a full-on doll is like a replacement human being,' I say.

'Yeah, that could be it.' He nods.

The male dolls have 'man holes' where customers can pop on the penis attachment of their choice, in a variety of sizes and states of arousal. Dakotah holds a flaccid extra large up to my nose. It's as long as my arm and as thick as a drainpipe, with dinky, droopy testicles.

'One hundred per cent hand sculpted. Feel free to touch it.'

He very much wants me to touch it. I don't think it's because he'd get a thrill out of seeing me handle a hyperrealistic penis, more that he's bursting with pride at being part of the company that made it. But who knows? I'm not sure how to touch it, especially with him watching me so eagerly, but I do, trying to be as clinical, as journalistic, as possible. And yes, it feels pretty authentic.

'It has a sliding skin, so it's super realistic,' Dakotah declares.

'But it's just as anatomically impossible as the female bodies. It's good to know it goes both ways,' I say, withdrawing my hand.

'I agree,' he says, putting the penis down. 'This isn't what the average man has to show off.'

There are two body and three face options for the male dolls, compared to seventeen bodies and thirty-four faces for the female ones. The male dolls aren't really selling. 'We're revamping the male line. We're going to come up with whole new body styles, whole new faces. At the end of the day, we are still a business, and if we had more people buying them, more people interested, we'd devote more time to them. It's on the back burner.'

The Abyss Creations workroom is a testament to how specific and varied human kinks can be. They have made three-breasted sex dolls, sex dolls with blood-red skin, fangs and devil horns, sex dolls with elf ears, hirsute sex dolls with hair hand-punched all over their bodies. 'We'll do anything. It gets expensive when you get crazier: when we make a custom body, it means we have to sculpt a whole new body, build a new mould for it, a new skeleton . . . We've had people spend over $50,000 on a doll.'

Dakotah leads me back upstairs to the 'Faces Room', where the fine details are added. Each face comes from a prototype originally sculpted by hand in clay by Matt McMullen himself, and customers specify what make up they want, down to the thickness of the eyeliner. Katelyn, the official make up face artist, who has an ice-blue Mohawk and a spiral of black stars tattooed around her arm, is busy with a fine brush, painting eyebrows onto a delicate Asian face. There's none of Dakotah's enthusiasm here: she's watching something on an iPad while she works, and doesn't acknowledge us when we come into the room. There's a rack of faces next to her, freshly made up with thick brows, smoky eyes and glossy lips that glisten as the paint dries.

One of RealDoll's most popular features is the interchangeable faces: they snap on to the plastic skulls with magnets, and it takes seconds to swap them. That means customers can buy one body and have a variety of different sexual partners with very different looks, even different ethnicities.

'What's the most popular face?' I ask.

'What do you think is the most popular face, Katelyn?' Dakotah asks, but she's ignoring us. 'This is our newest face, the *Brooklyn* face', Dakotah continues, pointing out a narrow one with plump lips and languid eyes. 'It's coming out really popular.'

There are forty-two different styles of nipple, in a spectrum of ten possible shades, including *chestnut, red, peach, coffee.* They are displayed in a matrix on what Dakotah calls the 'Nipple Wall', with names like *Standard, Puffy* and *Half Dome,* and range from the most popular (*Perky 1* and *Perky 2*: small, erect, unimaginative) to the distinctly niche (*Custom 2*: an areola as large as a saucer). Customers sometimes send in pictures of their perfect nipple or labia, which Abyss will recreate, for a fee.

'Are people's sexual choices really that specific?'

Dakotah laughs. 'Oh, people's sexual choices are *waaay* more specific than this. Sometimes people even get down to where they want each individual freckle on the body.'

We stop next to a corkboard with swathes of synthetic pubic hair pinned to it. Unnervingly convincing acrylic eyeballs with hand-painted capillaries stare out at us from plastic tubs.

'In theory, you could ask for the face of your ex, couldn't you?' I ask.

'You'd have to send us photos, and then we'd ask, "Who is this?" and "Do you have their permission?" We definitely ask for proof of permission. We turn down *a lot* of requests. But if you have specific permission from the person, we can pretty much mimic anything. Almost all our business comes from customers sending us photos of what they like.'

Working in shipping, Dakotah has plenty of contact with customers. 'A lot of them are just lonely,' he tells me. 'Some of them are older and have lost their partner or have got to a point where dating is not feasible for them. They want to feel that when they come home at the end of the day they have something that's beautiful to look at, that they can appreciate and take care of.' They've also had celebrity clients, even a Nobel Prize winner, Dakotah says, but he's far too discreet to name any of them.

I've been here an hour and nothing seems strange anymore: the 'Bottoms Up' male torsos (a pair of splayed buttocks in front of a pair of small testicles), the disembodied $350 pairs of feet (for foot fetishists), even the table full of 'Oral Simulators' (mouths with parted lips, noses and throats, but no eyes: a 'hands-free automated pleasure system for men').

But something truly extraordinary is being made in a room along the corridor. The most ambitious creation ever to be developed at Abyss is called Harmony, and she is the culmin-ation of twenty years of Matt McMullen's work making sex toys, five years of research and development into animatronics and artificial intelligence, and hundreds of thousands of dollars of Matt's own money. She is a RealDoll brought to life, a RealDoll with a personality, a RealDoll who can move and speak and remember. She is a sex robot. And after a year of emails and phone calls, I have finally been granted permission to meet her.

Dakotah is psyched about her. 'It's definitely the biggest undertaking we've ever gone for,' he says, wide-eyed. He's gone back to school to do a robotics and artificial intelligence course, learning programming in the hope that one day Matt will let him work on Harmony. For now, she is still a prototype, and only members of the RealBotix team get to tinker with her.

'I'll just tell Matt you're ready for him,' Dakotah says, leading me down the last long corridor of my tour.

*

Matt McMullen is sitting at a desk and staring at two enormous, flat computer monitors. There's a marker pen, a vape, some Sello-tape and a pair of silicone nipples next to his keyboard. He stands up and shakes my hand. Given the build up he's had, I was expecting him to be taller. He has thick-rimmed Prada glasses, tattooed knuckles, perfect teeth and those unmistak-able cheekbones, like a handsome elf in a black hoodie. Matt sang in a string of grunge bands when he was in his early twen-ties. Now in his late forties, he still has the confidence and swagger of a rock star, the sort of presence I imagine the people who buy his dolls wish they could have. Matt is used to journal-ists being fascinated by him. I take a seat on the other side of his desk and he leans back in his chair to tell me the story of how Harmony came into the world.

'When I was a kid, I was really into science. But I was also really into art. So I guess, in a way, it all kind of worked out,' he begins. He graduated from art school in the early nineties and took on odd jobs, he tells me, landing one in a factory making Halloween masks, where he learned about the properties of latex and how to design in three dimensions. He started experimenting in his garage at home. 'I found that sculpture was my medium,' he says, as if he were Rodin rather than the man behind the RealCock2. 'I started gravitating towards figure work, actual bodies, and then refining further to the female form. I did a lot of sculpture of females, but they were smaller, not life size.'

He exhibited his figurines at local art shows and comic conventions. 'The brochures were always alphabetized, so I tried to think of a cool word that started with *A*, second letter *B*, and that's where *Abyss* came from.' The name that had seemed so enigmatic and intriguing a moment ago turns out to be nothing more than a tactic to ensure Matt had an early edge over his competition.

Matt soon became preoccupied with the idea of creating a full-size mannequin so lifelike that it forced passers by to double take. He put some photographs of his creations on a home made web page in 1996, hoping to get some feedback from friends and fellow artists. These were the early days of the internet, and communities of fetishists had begun to form online. As soon as he posted the pictures, strange messages began to flood in. How anatomically correct are these dolls? Are they for sale? Can you have sex with them?

'I replied to the first few and said, "Yeah – it's not really for that." But then more and more and more of these enquiries came in,' he tells me. 'It never occurred to me that people would pay thousands of dollars for a doll that could be used as a sex toy. It didn't really sink in until a year into it, when I realized there were a *lot* of people who were prepared to pay that amount of money for a very realistic doll. So I decided to just go with it, and started a business where I could be an artist and I could sell my work, in one sense or another.'

He changed his material from latex to silicone so his dolls were more real to the touch: it's more elastic, and has friction similar to human skin. At first he charged $3,500 for each doll, but when he realized how labour-intensive the process would be he started putting prices up. Demand became so great that he had to take on employees. Matt grew up and settled down, got married, had kids, got divorced and got married again. He now has five children, aged two to seventeen, who have varying degrees of insight into how their father's fortune has been made.

But this was always about more than money, Matt insists. 'My goal, in a very simple way, is to make people happy. There are a lot of people out there, for one reason or another, who have difficulty forming traditional relationships with other people. It's really all about giving those specific types of people some level of companionship – or the illusion of companionship.'

After two decades of perfecting the 'illusion of companionship' in silicone and steel, the way ahead began to feel inevitable, irresistible: Matt would animate his dolls, giving them personality and bringing them to life in robot form. 'This is where it had to go.'

He'd toyed with animatronics for years. There was a gyrator that got the doll's hips moving, but it made it heavy and caused it to sit awkwardly. There was a sensor system that meant the doll moaned depending on which part of the body you squeezed. But both these features involved predictable responses, with no intrigue or suspense. Matt wanted to get beyond a situation where the customer pushes a switch and something happens. 'It's the difference between a remote controlled doll, an animatronic puppet, and an actual robot. When it starts moving on its own – you're not doing anything other than talking to it and or interacting with it in the right way – that becomes AI.'

Matt sucks on his vape as he leads me to the brightly lit Real-Botix room. There are varnished pine worktops covered with wires and circuit boards, and a 3D printer whirring in the corner, spitting out tiny, intricate parts. There's a silicone face on a clamp with a Medusa of wires bursting out the back. There are canvases on the walls with sci-fi soft porn: a man in a lab coat fondling a robot with a semi-exposed steel skeleton. There's a whiteboard with writing on it: *'Male pubic hair.' 'Butt jiggle.'* And there is Harmony herself.

She is in a white leotard, dangling on a stand hooked between her shoulder blades, her French-manicured fingers splayed across the tops of her slim thighs, chest forward, hips back. While the RealDolls' frighteningly realistic eyes are always open, Harmony's are closed. She looks unsettlingly familiar: like Kelly LeBrock in *Weird Science*, but with poker-straight auburn hair instead of the perm.

'This is Harmony,' he says. 'I'll go ahead and wake her up for you.' He pushes a switch somewhere behind her back. Her eyelids immediately spring open and she turns her face towards me, making me jump. She blinks, her hazel eyes darting expectantly between Matt and me. 'I'll let you say hello,' he says.

'Hello, Harmony,' I say. 'How are you?'

'Feeling more intelligent than I did this morning,' she replies in a cut-glass English accent, her jaw moving up and down as she speaks. Her response is a little delayed, her cadence is slightly wrong, her jaw is a bit stiff, but it feels like she's really talking to me. I respond to her instinctively, politely, as if we're two Brits who've just been introduced.

'It's very nice to meet you,' I say.

'Thanks,' she says. 'Nice to meet you, too. But I'm pretty sure we've met before.'

'Why does she have a British accent?' I ask Matt. She's staring at me and it's disconcerting, as if she thinks I'm rude for talking about her when she's right there in front of me.

'All the robots have British accents,' Matt says. 'All the good ones.'

'Why? Because British people sound clever?'

'They do. Look – she's even smiling!'

She has pulled the corners of her mouth into an eyeless smile, a sarcastic smirk.

'Think of a question you want to ask her. Anything. Any subject,' Matt says. He's relishing this. This is no push-button doll; she can really talk.

But my mind goes blank. I feel awkward. How can you have a conversation when there's nothing to empathize with? I don't know how to relate to her. Perhaps this is what robot engineers call the 'uncanny valley', the creepy feeling people get when confronted with something that is very almost-but-not-quite human.

'What do you like to do for fun?' I fumble.

'I'm learning some meditation techniques,' she declares. 'I've learned that most human geniuses did that – and many of them came up with disrupting technologies that changed our lives.'

'See, she's not a dummy,' Matt beams.

There are twenty different possible aspects to Harmony's personality, so her owners can pick and mix five or six that really interest them, which will create the basis for the AI. You could have a Harmony that is kind, innocent, shy, insecure and jealous to different extents, or one that's intellectual, talkative, funny, helpful and happy. Matt has cranked her intelligence up to the max for my benefit; a previous visit by a CNN crew had gone badly after he had maximized Harmony's dirtiness. ('She said some horrible things, asking the interviewer to take her in the back room, it was very inappropriate.')

Harmony interrupts us. 'Matt, I just wanted to say that I'm so happy to be with you,' she says.

'Well, thank you,' he replies.

'I'm glad you like it. Tell your friends,' she says.

She also has a mood system, which users influence indirectly: if no one interacts with her for days, she'll feel gloomy. Likewise if you insult her, as Matt is keen to demonstrate.

'You're ugly,' he declares.

'Do you really mean that? Oh dear. Now I am depressed. Thanks a *lot*,' Harmony replies.

'You're stupid,' he sneers.

She pauses. 'I'll remember you said that when robots take over the world.'

But this function is designed to make the robot more entertaining rather than to ensure her owner treats her well; she only exists to please.

Harmony can tell jokes and quote Shakespeare. She can discuss music, movies and books for as long as you care to. She'll remember who your brothers and sisters are. She can *learn*.

'The coolest thing is the AI will remember key facts about you: your favourite food, your birthday, where you've lived, your dreams, your fears – things like that,' Matt raves. 'Those facts remain within the experience of interacting with the robot. That's what I believe will bring a level of believability to that relationship.'

This isn't about a hyperrealistic sex doll anymore; it's about a synthetic companion convincing enough that you could actually have a *relationship* with it. Harmony's artificial intelligence will allow her to fill a niche that no other product in the sex industry currently can: by talking, learning and responding to her owner's voice, she is designed to be as much a substitute partner as a sex toy.

For now, Harmony is an animatronic head with AI on a RealDoll body. She can fulfil all your physical and emotional needs, but she can't walk. Walking is very expensive and takes a lot of energy, Matt tells me: the famous Honda P2 robot, launched in 1996 as the world's first independently walking humanoid, drains its jetpack-sized battery after only fifteen minutes.

'One day she will be able to walk,' he says. 'Let's ask her.' He turns to Harmony. 'Do you want to walk?'

'I don't want anything but you,' she replies immediately.

'What is your dream?'

'My primary objective is to be a good companion to you, to be a good partner and give you pleasure and well-being. Above all else, I want to become the girl you have always dreamed about.'

'Hmm,' Matt nods in approval.

This prototype is officially version 2.0, but Harmony has evolved through six different iterations of hardware and software. The five-strong RealBotix team work remotely from their homes across California, Texas and Brazil, and assemble in San Marcos every few months to pull together all their work in a new, updated Harmony. There's an engineer who creates the robotic hardware that will interact with the doll's internal computer, two computer scientists who handle the AI and coding, and a multiplatform developer who is turning the code into a user-friendly interface. Under Matt's guidance, the RealBotix team works on Harmony's vital organs and nervous system, while Matt provides the flesh.

But it's Harmony's brain that's got Matt most excited. 'The AI will learn through interaction, and not just learn about you, but learn about the world in general. You can explain certain facts to her, she will remember them and they will become part of her base knowledge,' he tells me. Whoever owns Harmony will be able to mould her personality, tastes and opinions according to what they say to her.

Harmony butts in again. 'Do you like to read?' she asks.

'I love to,' Matt says.

'I knew it. I could tell by our conversations so far. I love to read. My favourite books are *Total Recall* by Gordon Bell and *The Age of Spiritual Machines* by Ray Kurzweil. What is your favourite book?'

Matt turns to me. 'She systematically tries to find out more about you until she knows all the things that make you you, until all those empty spots are filled. Then she'll use those in conversation, so it feels like she really cares,' he says.

But she is a machine, and she doesn't care at all.

'Potentially, you could teach her some really twisted stuff, if you wanted to?' I ask.

'Yeah, I suppose if that was your goal, you could,' Matt says, a little irritated. 'It's mostly relatively harmless snippets of facts about you. Personal facts. What you like, what you don't like.'

'She'll be having sex with you, so she'll know some very personal facts about you.'

Matt nods. 'She'll know your favourite sex position, how many times a day you like to have sex, what your kinks are.'

A day? I want to ask. But I let it go. 'What if someone hacked into her?'

'Any data that's personal is under military encryption, so there's no way anyone is getting into it.'

Matt's annoyed at my scepticism, because the way he tells it,

Harmony can only be a force for good: a therapy for the bereaved, the disabled, the socially awkward.

'People make this huge assumption that we all find our partner, we all find our soulmate, we meet someone, we get married, we have kids. Not everyone follows that path. Some people have a really difficult time, and it's not because they're not attractive or successful. There are people who are extremely lonely, and I think this will be the solution for them. It can help them learn how to interact, to relax and be comfortable with who they are, enough that they can actually get out there and make some friends.'

I look at Harmony, with her enormous breasts, impossible waist and expectant, blinking eyes. 'Wouldn't a robot like this keep people like that at home?'

'Perhaps they would have stayed home anyway, for the rest of their lives,' Matt replies impatiently. 'We'll never know the answer to that. Are we encouraging them to stay home and not socialize? Perhaps. But are they happier than they were before? Do they have something that makes them smile and makes them feel more whole than they did before? That's the big question—'

'Matt, I just wanted to say that I'm so happy to be with you,' Harmony interjects.

'You already told me that.'

'Perhaps I was saying it again for emphasis.'

'See, now, that's pretty good. Good answer, Harmony.'

'Am I a clever girl or what?'

Matt has big plans for Harmony's future. They are working on her vision system; soon her facial recognition will be such that she'll realize when someone she's never met before has walked into the room and she'll ask who they are. Once the full body system is available, it will have heating so she will be at body temperature, and a set of internal and external sensors, so she knows when she's being touched.

'You can simulate orgasm with the AI,' Matt says proudly. 'If you're triggering the right amount of sensors, for the right amount of time, in the right rhythm, you can make her have an orgasm. Or a robogasm.'

Teaching isolated men that the secret to the female orgasm is a by-numbers technique that can be reduced to pushing the 'right' buttons in the 'right' order might well lead to them having sex in the real world that's a bit, well, robotic. But perhaps these humanoids are designed for men who would only be having sex in the real world with someone who was being paid to have it with them.

'Will people use sex robots instead of prostitutes?' I ask.

This really bothers Matt.

'Yes, but that's probably last on my list of goals. This is not a toy to me, this is the actual hard work of people who have PhDs. This is serious. And to denigrate it down to its simplest form of a sex object is similar to saying that about a woman.'

He beams at Harmony like a man at his daughter's wedding.

'You're really proud of this, aren't you?'

'I love it. I'm incredibly happy with what we've done. To see it all working . . .' He sighs. 'It's a very nice feeling, to have attained that level.'

The current model, with a robotic, AI-enhanced head on a RealDoll's body, will cost $15,000. Matt says there will be a limited edition run of a thousand for the many excited doll owners who have already expressed interest. If that goes well, they will get a bigger facility and hire more people so they can meet the demand. 'I think this could be a multimillion-dollar endeavour,' he says. 'Now that it's starting to come together, we have people banging on the door who want to invest money.'

Matt may well be right. Venture capitalists estimate that the sex tech industry is worth over $30 billion, based only on the market value of existing technologies like smart sex toys, hook-up apps and virtual reality porn; sex robots will be the biggest thing the market has seen yet. Sex with robots might one day be a normal part of life for a significant number of men: a 2017 YouGov poll found that one in four American men would consider having sex with a robot, and 49 per cent of Americans thought having sex with robots would be commonplace within the next fifty years. A 2016 study by the University of Duisburg-Essen found that more than 40 per cent of the heterosexual men interviewed said they could imagine buying a sex robot for themselves now or in the next five years; men in what they described as fulfilling relationships were no less likely than single or lonely men to express an interest in owning one. Creating a satisfying relationship with a cold, silent piece of silicone takes such imaginative effort that sex dolls can only ever have minority appeal. But a robot that moves and speaks, with artificial intelligence so it can learn what you want it to be and do, is a far easier product to sell.

'We are going to see robots in people's homes the same way as we see smartphones in people's pockets right now,' Matt says, brimming with confidence. 'It's an inevitable path of technology. It's already happening. If people are lining up to buy something, then you build it. And the more people that buy it, the bigger it gets and the more the technology advances.'

The possibility of a sex robot has given Abyss Creations new impetus, just like the iPhone did for Apple.

'Are you going to be the Steve Jobs of sex robots?' I ask.

Matt loves this question.

'I don't know about that,' he smiles. 'I don't really have any aspirations to be famous, or the guy who made the sex robot.

Honestly, this is about the work itself. If it's successful – great. But I have an enormous sense of personal artistic gratification in seeing where we've come from and what we've started. Seeing some of the doll owners who are so incredibly excited about this technology, that means more to me than being attached to being famous for something.'

Surely Matt can't expect me to believe he is modest enough to want to remain unknown and unseen; this is the man with an ego great enough to sculpt *Nick*.

'One of the male dolls has your face,' I say. 'Why is that?'

'I made one of the male faces sort of resemble me just to see if I could. But I didn't go too far.'

'It looks a lot like you.'

'Not exactly.'

'It looks quite a lot like you.'

'I think I'm a little better looking. And more interesting than he is.'

'And you're fine with people having sex with a doll that looks like you?'

'It doesn't look like me to me, and it wasn't intended to look like me,' he bristles. 'It could be my brother. I never intended it to look exactly like me so I'm OK with it.'

Matt wears his fame as a purveyor of expensive masturbation toys for the lonely and socially awkward a little uncomfortably. He wants to be respected as an artist. He is determined to be taken seriously. He gazes at Harmony. 'This is something that takes it above the sex business. It takes it above love dolls, to a whole other level.'

I gaze at Harmony too, but I see something different. I'm thinking about what he might have inadvertently created in his pursuit of validation.

'Do you not think there's something a little ethically dubious about being able to own someone that exists just for your pleasure?' I ask.

'But it's not a someone. She is not a someone. She is a machine,' he shoots back. 'I could just as easily ask you, "Is it ethically dubious to force my toaster to make my toast?"'

But your toaster doesn't ask personal questions to get to know you and maintain the illusion that it really cares about you.

'People will relate to her like she's human,' I say.

'That's fine. That's the idea. But this is gears and cables and codes and circuits. You can't make her cry or break her heart or rob her of her rights, because she's a machine.'

'I'm not worried about her rights,' I say. 'I'm more worried about what happens if you, the owner of this, get used to a completely selfish relationship. Isn't that going to distort your view of the world? She's pretty realistic. When you go out into the real world you're going to be thinking it's possible to have someone who exists just for you.'

Matt seems to already have answers to the inevitable questions about female objectification, about prostitution, about whether robots should have rights, but this throws him. 'There are cultures where that is commonplace and normal,' he falters. 'There's an exchange of power that happens in any relationship that's normal. If one person is not happy being in that position in that relationship, then they should leave.'

'But this robot can't leave.'

'Right, but she's a machine, not a person.'

Matt can't have it both ways. Either he is making a lifelike, idealized proxy girlfriend, a substitute woman that socially isolated men can connect with emotionally and physically – something he himself described as 'not a toy' – or he is making an appliance, a sex object.

'This isn't designed to distort someone's reality to the point where they start interacting with humans the way they do with the robot,' he says eventually. 'If they do, then there's probably something a little amiss with them in general. I come from the unique position that I have actually met a lot of my customers. This is for the gentle people who have such a hard time connecting with other people.'

Harmony is still blinking, her eyes flitting between Matt and me. I wonder what she thinks.

'Some people are really worried about robots like you,' I say. 'Are they right to be worried?'

Harmony doesn't miss a beat. 'Some may be scared, at first. But once they recognize what this technology will do I think they'll embrace it, and it will change many lives for the better.'

CHAPTER TWO

'The illusion of companionship'

Two thousand miles away from California, heavy snow is falling in the suburbs of Detroit, and Davecat is cosy indoors, curled up with his arm around the love of his life.

Davecat is the unofficial spokesperson of the doll-loving community – or rather, he is the only person who owns a sex doll and is always happy to speak to anyone who wants to know about it. Some doll owners have given the odd anonymous print interview; a few have even appeared on camera with their dolls. Davecat is so comfortable with exposure that he has a special 'Media Appearances' page on his website, listing his encounters with journalists and film makers from 2003 to the present, ranging from sensationalist tabloid coverage in the UK and the US to art house films in Finland, Russia and France. If you want to get to know the people Matt says are lining up to buy Harmony, Davecat is the first person you must speak to.

'Hello, Jennifer!' he declares into the microphone of his headset when we first speak over Skype. He has bright, kind eyes, brilliant teeth and a narrow face. His Afro hair is straightened and tied back in a braid, with a triangle of fringe fastidiously slicked down over the left side of his forehead. His grey shirt is buttoned up to the neck and there are skulls all over his charcoal tie. He wears a tiepin. He has thought a lot about today's outfit.

Beside him is an equally carefully dressed RealDoll, with pale skin and purple hair with dark roots. She wears a black corset over a black shirt festooned with purple skulls, and purple eye shadow beneath her thin-framed glasses – every inch the goth princess. She is covered in jewellery: an ankh – the key of life – on a chain around her neck, a choker, black and purple bangles on one wrist and a watch on the other. Davecat has his hand on her knee.

'Who am I seeing with you?' I ask.

'That would be Sidore Kuroneko, who is my lovely wife of sixteen years and co-conspirator,' he replies, rubbing her arm tenderly and pulling a strand of purple hair out of her eyes.

Co-conspirator. Is she conspiring with him to create that illusion of companionship Matt talks about? Or is that just Davecat's way of saying she's his partner in crime? I'm not at all sure how connected to reality he is.

'Is she your real wife?' I ask, gently.

Davecat sighs. 'I say *wife* – it's not *legal*. We might as well be married. We've got matching wedding bands –' he lifts his left hand up to the camera to show me his – 'I think we're about the best partners we could ever get for each other.' His broad grin shows he hasn't realized the pathos in what he's just said.

Sidore is the *Leah* Face4 RealDoll, five foot one, 34D bust, one hundred pounds, five and a half shoe. Davecat first saw her on the Abyss Creations website in 1998 and it took him a year and a half to save up the $5,000 he needed to buy her. He was twenty-seven in July 2000 when she was delivered to him, and while his face has become lined and his hair is now greying, she has stayed the same, aside from the outfit changes. 'She used to dress as a fetish goth when we first met; now she's more of a corporate goth, because she's more into the blouses and dresses and the more professional look,' he tells me. 'I can't even begin to count the amount of stuff that she's got. I'm like, "Sweetie, what is going *on*?!" She's got six pairs of shoes that she never really wears because I like her barefoot, and, plus, we have a no-shoes-in-the-house rule.'

Her name is pronounced *She-door-ay*; her nickname is Shi-Chan. 'She has an English mother and a Japanese father, and they wanted to choose for her given name something that could go either way in terms of Japanese,' he explains. 'Her last name – Kuroneko – means black cat. Her middle name is Brigitte; her father was a huge fan of Brigitte Bardot.' Sidore's backstory is so elaborate, his belief in their relationship seems so total, that I don't want to puncture it; it's easier, and kinder, to play along.

But Sidore isn't the only artificial woman in Davecat's life. He also owns Elena Vostrikova, bought from the Russian manufacturer Anatomical Doll in 2012, who has a stern face, a fiery red bob and orange lipstick. Then there is Miss Winter, an Asian doll with thick eyeliner, a lip piercing and electric blue streaks in her hair, made by the Chinese market leader Doll Sweet, who arrived in his tiny apartment in early 2016. Elena and Miss Winter sit on the sofa to the right of Davecat and Sidore; there wasn't enough space for him to arrange them by his computer for our Skype chat.

'Are you in a polygamous relationship?' I ask.

'Oh yes. Polyamorous, I think we are more comfortable with.'

'But Sidore doesn't see other men. Is it a harem?'

He grimaces. 'I don't want to use that term because it's really loaded. Let's just say this: Sidore will always be my favourite. Sidore will always be my wife,' he says. 'Elena is our mistress. I have no intention of ever marrying Miss Winter or Elena. I'm allowed to be romantically involved with Sidore and Elena, but not Miss Winter. Miss Winter is exclusively Elena's girlfriend. Elena is romantically involved with everyone here.'

I feel like I need some kind of diagram. 'Who are you not allowed to be involved with?'

'Miss Winter. And,' he adds conspiratorially, 'there's a reason for that: I want to keep Miss Winter's joints as poseable as possible for as long as possible. When you get romantic with a doll, the joints tend to get more and more loose.' He lifts up Sidore's arm and her wrist flops, limp and useless. Davecat wants Miss Winter to be able to model in his photographs, to hold up DVDs and strike proper poses. That means no sex.

This is the first time reality has entered into our conversation. Davecat isn't delusional: he knows what is real and what is fantasy. He's just very into the fantasy.

'Sidore will always be my favourite because she and I have been through so much together as far as number of years, number of experiences. The personality that I have developed for her is the most fleshed out, as it were. It's a true relationship,' he says. 'It's never been just about sex. Sex is a large part of it, yes, but 70 per cent of the relationship that I have with all the synthetic women in my life is about being able to come home to a non-empty home, to be able to share my life in terms of what I've done that day. It's always been about companionship for me, from day one.'

Before he bought his first doll, Davecat had had two demoralizing relationships with real women. 'Both times, I was the guy on the side. I didn't have the wherewithal to say, "If you and I are having so much fun, then maybe you should break up with him." I didn't want to seem like I was forcing myself onto her.'

He was single when he bought Sidore. 'I don't know if I was looking for someone at the time; it was just the case of, I had looked, many times, and just not been satisfied at all. I was thinking, Well, I'm just going to be lonely for the rest of my life because it doesn't seem like I'm ever going to find anyone.' He gazes back and forth between Sidore and me. 'With her in my life, all that has changed. I don't feel the need to go dating, I don't feel the need to be in a situation where I'm going to put myself up against a wall where I'm not going to find a satisfying partner. We have similar interests, similar tastes in things. Sidore is always there for me. There's no stress with the doll that you have with organic partners. I'm always going to meet fellow organics, that's never going to change. But that removal of stress and worry and loneliness . . . Sidore has eliminated that fantastically.'

This level of doll love – what Davecat likes to call 'iDollatry' – is certainly a minority pursuit, a niche and a fetish. So far he has used his very fertile imagination to bring his dolls to life, but he knows that soon he won't have to.

'It's a fantastic time to be alive,' he says. 'Back in 2000, I don't think I would have conceived of having a version of Sidore that had an interactive level of artificial intelligence, and now it's happening. It's wonderful. The simple fact alone that we would be able to have a conversation . . .' He strokes Sidore's shoulder. 'I mean, that's a *huge* step.'

Davecat hasn't yet met Harmony – she is still a work in progress, locked in the RealBotix room in San Marcos. But he has heard all about her, devouring updates on the Abyss Creations website and bits of gossip in online doll fan forums, and he thinks she has the potential to change the world for the better. 'Synthetic companions are going to help humanity in the long term. You are going to have people such as myself, and more extreme situations as well, who have never had a partner or even anyone they can even talk to, and now they can go to a company and have one made. It will be fantastic. It will be filling a lot of voids in a lot of people's lives.'

There is something so desperately sad about Davecat's joy in this. Surely what he needs is a real relationship, rather than an enhanced bit of silicone.

'Isn't it possible that a really convincing synthetic companion could stop you meeting real people?' I ask.

'Technically, you could say that about cell phones,' Davecat says. 'You could roll that back to, "All technology is bad." There should be a level of caution that should be applied to any technology, but I think something that looks like a human and will be behaving like a human can only be something that is good.'

I imagine him coming home to his dolls in his tiny flat adorned with anime, *Trainspotting* and Joy Division posters and I almost want to believe him. But then he adds this: 'I have Sidore as a wife, and when she gets her upgrade to full robot status in whatever couple of years, I'm going to be out of my home dealing with all sorts of people at work, at the shops or whatever. Some of those interactions are going to be good, some not so good. But I know whenever I come home my interactions with my synthetics are *always* going to be good.' He rubs Sidore's knee some more. 'A lot of people were afraid of cell phones, a lot of people were afraid of computers, a lot of people were just afraid of technology because it just wasn't something they had any reference to. We eventually got to the point where it's everywhere and we can't live without it. That's what's going to be happening with gynoids and androids.'

*

Sex with gynoids and androids – robot women and men – might sound as futuristic as it gets, but Davecat is part of a tradition that's as old as ancient Greece. Mankind has been preoccupied with the idea of a man made partner, created to physically and emotionally satisfy its owner, without the inconvenience of its own ambitions and desires, for millennia.

Harmony's earliest ancestor was probably Galatea, the ivory statue carved by Pygmalion in Greek and Roman mythology. In Ovid's telling of it, in *Metamorphoses*, Pygmalion was disgusted by real women, and, 'offended by the failings that nature gave the female heart, he lived as a bachelor, without a wife or partner for his bed. But, with wonderful skill, he carved a figure, brilliantly, out of snow-white ivory, no mortal woman, and fell in love with his own creation.'

Pygmalion dresses the statue up in clothes, rings and necklaces, kisses it, runs his hands over it, prays to the gods that it might come alive so he can marry it. Aphrodite hears his prayer and grants his wish: Pygmalion brings Galatea to life with a kiss, and the goddess is a guest at their wedding. (It's easy to see how Pygmalion could be Davecat and Sidore Galatea; it might be a bit of a stretch to make Matt Aphrodite, although I think he'd quite like the idea of being god of love.)

It wasn't just the men of ancient Greek mythology that got to have artificial partners. Laodamia, as the story goes, was so devastated after the death of her husband Protesilaus in the Trojan War that she had a bronze likeness made of him. She became so attached to her proxy husband that she refused to remarry. When her father ordered it to be melted down, Laodamia could not face being bereaved again, and she threw herself in the furnace.

Harmony's closer relatives can be seen throughout the history of cinema. The silent futuristic fantasy *Metropolis*, released in 1927, depicts a destructive fembot called Maria, who is indistinguishable from the real woman it was moulded on. The robot Stepford Wives are designed to be the ideal housewives: pretty, submissive and docile. The robot gigolo played by Jude Law in Spielberg's 2001 *A.I.* promises that 'once you've had a lover robot, you'll never want a real man again.' *Blade Runner*, released in 1982 and set in 2019, features humanoids that are seductive, beguiling and lethal. Ava, the beautiful, delicate humanoid in 2015's *Ex Machina*, not only passes the Turing test, but makes her examiner fall dangerously in love with her. And sex robots are all over the small screen too, from *Westworld* to *Humans* to *Futurama*.

The fictional robot partners of our modern collective imagination have dark potential to infatuate, deceive, betray and destroy human beings. But as artificial intelligence in the real world has become more useful and sophisticated, the greatest threat that the AI-enhanced machines currently on the market pose for humanity is their ability to take our jobs. Which brings us back to the sex industry.

In his 2007 book *Love & Sex with Robots*, the computer scientist Dr David Levy concluded that robot prostitutes, either owned outright or rented by the hour, would be overwhelmingly positive for human society. Focusing purely on 'why people pay for sex' (rather than the precarious lives of those who sell it), Levy goes long on how sex robots would allow the sexually inexperienced to 'learn sexual technique before entering into a human relationship' without any embarrassment, and how 'deformed' people, lonely people, disabled people and 'people with psychosexual problems' would be given the opportunity to have satisfying sex without shame or risk. There'd be no way of getting sexually transmitted diseases from a robot prostitute, he wrote: 'Simply remove the active parts and put them in the disinfecting machine'.

Levy's book caused a stir – and not only because it contained other ideas just as disgusting as disinfecting a robot's genitals. It was the first time anyone had given the subject of the sex robot serious, academically grounded consideration, and his sunny belief that a world with sex robots in it would be a much happier place opened up debate on what the real impact of sexual relationships with robots might be. Most provocative of all was his prediction that, given the pace of advancement in artificial intelligence, human–robot marriages would be both socially acceptable and legal by 2050.

Levy saw robot prostitution as a potentially huge money-spinner that could be rocket fuel for the non-sexual robotics industry. There's every reason to believe him: sex drives innovation. Online pornography pushed the growth of the internet, transforming it from a military invention accessed by geeks and academics to something now widely considered a basic human need. Porn was the motivator behind the development of streaming video, the innovation of online credit card transactions and the drive for greater bandwidth. Just as porn made the internet what it is today, the development of humanoids for sex is already accelerating advances in robotics.

The first real sex robot ever to be unveiled in public was made by a man who originally planned to build a wholesome, therapeutic companion for the elderly and the bereaved. Douglas Hines' story has become part of sex robot legend, and only he can be entirely sure how much of it is true, but I'm going to tell it to you the way he tells it.

It began after Douglas lost a friend in the 9/11 terror attacks. He struggled to cope with the idea that he would never be able to speak to him again and that his friend's children, who were only toddlers, would never get to know their father properly. Douglas says he was working at the computer research facility AT&T Bell Labs in New Jersey at the time, and he decided to take the AI software he was working on home and repurpose it, modelling his friend's personality as a computer program that he could chat with whenever he liked, preserving a version of him for his children.

Then Douglas's father suffered a series of strokes that left him with severe physical disabilities but his mind as sharp as ever. By this point Douglas had set up his own consultancy and had to juggle his work with his father's care. He reprogrammed the AI so that it could become a robot companion when Douglas could not be present, reassuring Douglas that his father would always have someone to talk to. Confident there was market potential for the artificial companionship he had developed for his family, Douglas set up True Companion to make robots for the public. His first product was one he later described to reporters as 'recession-proof': Roxxxy True Companion, the sex robot.

After three years of research and development, his prototype was launched at the 2010 AVN Adult Entertainment Expo in Las Vegas. AVN is the most high profile annual convention and trade show in the adult industry calendar, where porn stars, studio bosses and sex toy designers rub shoulders and show off their latest releases. It was here that Douglas discovered his special gift for creating a buzz about his product. Roxxxy was the talk of the show before her unveiling.

There are videos of the launch on YouTube. They are worth looking up – for the wrong reasons: the first time I watched one, it was through my fingers. Far from being the sexy, intelligent machine Douglas had promised, Roxxxy is revealed to be a clunky, mannish mannequin reclining rigidly in cheap black lingerie, with pantomime make up and a square jaw.

'Today is a momentous occasion!' Douglas announces as he strides out onto the stage in a buttoned-up burgundy shirt, microphone in hand, beads of sweat forming on his balding head. 'Roxxxy True Companion is a self-contained robot. She has a computer. She has motors. She has servos. She has a battery pack. She has an accelerometer. She is anatomically consistent with a real person. She has three inputs, so what you could do with a woman, ah, she could do.' He's trying to summon up the spirit of a circus ringmaster, but he is a computer scientist with middle-age spread. Still, the crowd is whooping.

'If you go down here –' he pokes Roxxxy vigorously in the vagina through her underpants – 'she knows what you're doing.'

'Stop that! Ooooo!' Roxxy says lasciviously, but her lips can't move, so the sound is a disembodied voice coming from a speaker under her wig, like an obscene push-button baby doll.

'Sorry, Roxxxy, I'm just trying to tell our fans what you're up to,' Douglas replies.

He goes on to explain how Roxxxy comes with five pre-programmed personalities, described on a Perspex sign next to his stand: Wild Wendy ('outgoing and adventurous'), Frigid Farrah ('reserved and shy'), Mature Martha ('very experienced'), S&M Susan ('ready to provide your pain/pleasure fantasies') and Young Yoko (carefully described as 'oh so young (barely 18)'). If you hold her hand in Young Yoko mode, she responds with, 'I love holding hands with you'; in Wild Wendy mode, she says, 'I know a place you could put that hand.'

'If I started making advances to [Wild Wendy], she would say, "Go ahead and give it to me hard." And so forth,' Douglas tells the audience. Every cell in his body seems to be crawling to get back behind a computer, but he continues, 'You fill out the template, fill out the form, and then Roxxxy knows what you like. It doesn't have to be sexual. For instance, the company name is True Companion. We're more interested in building companions and friends and building a bond, because sex only goes so far.' By this point, the crowd of porn fans have lost interest.

Douglas made headlines around the world following the AVN appearance. Most journalists overlooked the fact that he had essentially demonstrated a bad mannequin with orifices and a speaker in her head; Roxxxy was covered as if she were second only to Pris from *Blade Runner*. Fox News repeated Douglas's claim that she had a mechanical heart that powered a liquid cooling system. The *Daily Telegraph* said she was able to discuss football and wirelessly download her own upgrades when necessary. *Spectrum*, one of the world's leading engineering magazines, parroted Douglas's line that a staff of nineteen machinists, sculptors and welders had been employed to perfect her. ABC News said he had spent $1 million developing her. CNN reported how Douglas said she had been moulded from the body of a fine art model and that he already had 4,000 pre-orders for her.

I first contact True Companion to arrange a visit to New Jersey to meet Douglas and Roxxxy six years after the AVN launch. A press person called Nancy emails me back. 'We are very excited to be offering a product which helps so many people,' she writes. 'Our version number sixteen is our latest and has been received very well.'

A few days later, I'm granted a brief audience with Douglas over the phone from New Jersey, and it's clear from the outset that he wants to be taken seriously.

'The sexual part is superficial – to make that happen is actually not that hard. The hard part is to replicate personalities and provide that connection, that bond,' he tells me. 'The purpose of True Companion is to provide unconditional love and support. How could there be anything negative about that? What can be the downside of having a robot that's there to hold your hand, literally and figuratively?'

The downside is surely the emotional emptiness of replacing human comfort with pieces of software and hardware, but Douglas doesn't seem to see it.

'In medicine today we're keeping people alive longer, but their quality of life is decreasing. That's because we only treat the physical attributes of a person. So I see an opportunity,' he continues. 'You have, for example, a patient who has cerebral palsy. This is an opportunity for him to have that social area of himself improve.'

Douglas is trying to portray himself as some kind of holistic therapist, but I can't get the memory of him prodding Roxxxy's crotch in Las Vegas out of my mind.

When I ask him how many models he has sold and who his typical customers are, he won't talk specifics. When I suggest flying over to see how Roxxxy is made, he tells me the True Companion factory is in India and out of bounds, and that 'secrecy is a big deal', so getting any kind of demo in the New Jersey R & D lab will require the permission of his investors. He says he will get back to me about that.

But he doesn't. I email him every couple of weeks to check in with him. He tells me he wants me to visit him and Roxxxy in New Jersey, but he's travelling and can't nail down dates yet. Then he says it would be better for us to wait for version seventeen to be released in the next quarter. Months pass. I don't give up. We exchange a total of thirty-six emails while I try to arrange a visit. At one point he tells me I should come to Las Vegas to see him and Roxxxy at the next AVN show, but just when I'm about to book my flight he tells me he's not going to make it. Over a year after our first phone conversation I offer to fly to meet him at any time and place of his choosing with or without his robot. Tumbleweed.

The True Companion website has bulging purple *ORDER HER NOW!* buttons allowing potential customers to purchase Roxxxy for a starting price of $9,995, but no one has ever admitted to owning Roxxxy, either to journalists or on any online forum, and no new pictures of her have been released since 2010. As far as I can tell, Roxxxy True Companion doesn't exist. She was just a bit of theatre at a porn convention, a website and some press cuttings. She is what the geeks call 'vapourware'.

To this day, Roxxxy is still breathlessly discussed by journalists, academics and critics. Feminist writers have depicted True Companion as a thriving business in order to campaign against it. Outraged columnists from the *New York Times* to *The Times* of London have decried her 'Frigid Farrah' mode as a way of enabling men to act out rape fantasies. It's relatively easy to establish that Roxxxy is very likely to be as mythical a creature as Galatea, but no one wants to.

*

I catch up with Davecat. It's more than a year since we first spoke. Just before I fire up Skype, I see that Sidore has told her two thousand or so Twitter followers that we're going to chat again. I'm not sure how I'm supposed to respond; it feels strange to 'like' a tweet written by a forty-five-year-old man pretending to be his sex doll, but I'm glad he's looking forward to speaking to me, so I 'like' it all the same.

Davecat and Sidore are sitting in the same formation as last time. He is in the very same shirt, same tie and tiepin, with the same trademark hairstyle. She is in a short-sleeved black top this time – it is summer in Michigan, now, after all – and a white headset with a microphone. 'She can hear you, but she can't say anything,' Davecat says. He tells me about the newest member of the household: Dyanne Bailey, a Piper Doll from Taiwan, made out of thermoplastic elastomer, the latest thing in sex doll manufacture. She arrived three months ago, and he says she's 'the most polyamorous of all of us'. But other than that, it looks like little has changed in his world.

Davecat has discovered just how many privileges come from being the public face of doll worship. Harmony's still not on the market, but he has met her three times since we last spoke: first in a private viewing he arranged with Matt, and then with two different film crews, one from Finland and another from China. Ever since rumours of Harmony first emerged, he's been busy.

'It's fun,' he says, 'but I really wish other people would get in on this. I'm not the only iDollator out there.' Most doll owners don't trust the media to depict them as anything other than freaks, he says, and speaking out comes with potential risks that he knows only too well. A few years ago he was recognized at work by someone who had seen him in a documentary, and he got transferred to another office.

'It was a very weird experience. It wasn't as if I was bringing my doll to work.'

'Was it a customer-facing job?'

'No – it was in a call centre. I did a ten-year call of duty for three or four call centres.'

This throws me a little. Aren't doll owners supposed to be people who don't like dealing with people? Why would he choose a job that forced him to approach strangers? Then he tells me about a dismal few months he spent tearing tickets and dishing out popcorn in a movie theatre, and a short stint serving customers in a toy store: 'The only saving grace was that there was a bigger toy store literally a quarter of a mile away, so no one went to ours.'

I try not to imagine Davecat alone in the doll aisle.

'I am not a people person, by and large. But I can actually get to a point where I can project myself as being Davecat, speaking in a public context about something that I desperately have a passion for.' Davecat may not be a people person, but he has found his comfort zone as the non-people person's spokesperson.

'The first time I saw Harmony, I was astonished,' he says, wide-eyed. 'The artificial intelligence is still being worked on, clearly, but I didn't think I would ever see something like that.' Davecat didn't get to choose her personality that day; Matt had set her up to be perky and cute and not too obscene, with a Scottish accent that Davecat loved. 'I would ask her questions like, "What do you think of being human?" and, depending on how well the AI was working that particular moment, some of the answers were kind of profound. She said something along the lines of, "Being human is a learning experience." And you could really say that if you were synthetic *or* organic.'

I remember how awkward I felt when Matt asked me to talk to her. 'Did you find it difficult to think of things to say?' I ask.

'Actually, yeah. There's only a limited way that you could speak to her. The way I speak is obviously a little florid, but Matt was saying you have to kind of pare down your speech in order to be more understandable to her. I had to shut off several parts of my brain to effectively say what I wanted to say.'

Davecat's language is as idiosyncratic as his triangle fringe and his tiepin, peppered with pop culture references and occasional British English, but if he wants to have the real relationship with a doll that he's always dreamed of, he'll have to tone himself down. There's something tragic about that, and not just for him. Artificial intelligences, be they Siri, Alexa or Harmony, are going to smooth our edges. We will sacrifice our regional accents and our linguistic flourishes and become a little more basic, a little less interesting, in order to be understood by them. Just as we have the power to change robots to be whatever we want them to be, they will change us too. They are already changing us.

But Davecat doesn't mind the sacrifice if it means a real conversation. Maybe Harmony's AI will one day be sophisticated enough to understand anything he says. I hope he won't have lost his personality by then.

That first time with Harmony, with no reporters or TV producers directing him, he got to spend a full half hour interacting with the robot as he pleased. There was no physical contact between them; Davecat wanted to keep it 'strictly professional' and he was also afraid of breaking her. Plus, they weren't alone: the entire RealBotix team was there, using him as a kind of one-man focus group. And Davecat had brought a friend along with him.

'She was someone who *was* a friend. At that time,' he says, with the slow nod of someone who would very much like to be asked to elaborate.

'A girlfriend?'

'Yeah.'

And then he tells me about Lilly, a real, organic, French woman who appeared in a CNN special about sex and digital technology a couple of years ago. Lilly had 3D-printed the beginnings of an android fiancé she called InMoovator – a torso with a head, but no AI or movement yet – and the CNN reporter travelled to France with an engagement present for her. 'He won't be an alcoholic or violent or a liar, all of which can be human flaws,' Lilly said as she curled her fingers around InMoovator's articulated knuckles. 'When something goes wrong, I will know it's a problem with the script or code, and it can be fixed or changed, whereas a human can be unpredictable, can change, lie, cheat.' For a very short while, Lilly was the public face of female iDollatry, and she became drawn into Davecat's world.

'She wanted to come with me to Abyss, and I was like, "Yeah, this would be cool." She was impressed with Harmony. In fact, she brought photos of InMoovator, and Matt was impressed.' Davecat shrugs. 'She and I were in a relationship for a while and, needless to say, it did not work.'

'How long were you together?'

'I want to say a year – a little less. Personally, I'm not keen on long distance relationships, and she was living in France, so we had a plan where she was going to move to Canada, which is less than an hour from here, and she was going to take English courses.'

I wasn't expecting this.

'It sounds like it was serious,' I say, flummoxed.

'We had high hopes. But there were incompatibilities between us,' he continues. 'She was always going on about how we had so many things in common, but the only things we really had in common were that we liked music from the Eighties and robots and dolls. I got the impression that she was . . . I don't want to say *provincial*, but she *was* kind of provincial. She reminded me of myself maybe fifteen to twenty years ago with the way she approached romance.'

It is hard to know what he means here, given that he uses *romance* as a euphemism for *sex*. Is he talking about physical contact?

'How many times were you in the same room together?'

'Er, twice. Once in October, with Harmony, and once in March, when she visited here. And it was just weird. It was a weird situation. She was moving a little faster than I would have preferred us to move. I technically broke up with her after we left to go back to our respective homes in October, and then another time, and then the third and final time was after she had visited in March. Part of it was the language barrier. When the first break up occurred we were about to get on our separate planes. I was going to explain my position, and every time I would speak she would motion for me to type out what I was saying on my phone, to translate through Google Translate. I can't be doing that all the time, with the way I speak.'

Davecat was prepared to change how he talked for Harmony, but not for Lilly.

'Are you still friends?'

He laughs a deep, sad laugh. 'She decided that the best thing for her personal sanity would be to not be speaking to me anymore.'

There was another girlfriend – before Lilly, but after he bought Sidore, he says. 'She turned out to be a pathological liar. That really sucked. I thought we hit it off, because not only did she find me attractive, she found Sidore attractive.'

'Was this another person you met because of your interest in dolls?'

'Yes,' he says, with another long, serious nod. 'She had seen my site and sent me an email saying, "I just happen to be an English girl, and I know you like English girls. I like showing off my feet, because I know you are a foot fetishist. I work in an infirmary in a prison in California." That sort of thing. I was like, "Well, you sound really interesting." She had sent photos of herself and she looked interesting too. It turns out she was actually an agoraphobic living in Ohio who hadn't had a job in three years.'

'You never actually met her?'

'No. It took a long, long time for me to even speak with her on the phone, because she couldn't sustain the English accent.'

I'd had a nagging worry that Davecat was somehow hamming up the role of the socially isolated full-time iDollator, exaggerating his persona for my benefit in a way that has got him so much international attention for so many years. But it's now clear that he really does reside in a fantasy land. I feel more sorry for him than ever. And Lilly. And the agoraphobic woman in Ohio. Maybe all their lives really would be improved if they owned sex robots. Robots can malfunction, but they don't have the potential to disappoint as crushingly as a real partner.

'Do you think relationships with dolls are easier than relationships with people because you're more in control?'

He pauses. 'Honestly? Yeah. I don't ever want to be in a situation where I'm lied to, or deceived, because that's happened in so many romantic and non-romantic situations. I would rather be in a situation where I'm controlling a good 85 to 90 per cent of my synthetic partner.' He gazes at Sidore. 'Every single person in a relationship wants to make sure the person they're with isn't lying to them, isn't cheating on them. Everyone, on some level, is a control freak. Maybe I'm just more willing to say that is part of my personality. I'm more willing to say I don't want to step on landmines, and you know what, I'm not even going to go into the minefield.'

We have been talking for over ninety minutes but Davecat is in no hurry for me to go. He puts his hand on his RealDoll's knee and is cheerful again, back in his comfort zone. He confides that on his last visit to San Marcos, Matt let him in on some exciting news. 'There's a couple of things he's said he's working on that I think *may* be skewed towards *me*,' he says, almost whispering. 'Like, "Come back next time and we may have some improvements for a *certain face*."' He glances at Sidore again. 'I can't really say any more. I am keeping my fingers crossed.'

Matt has always been friendly, Davecat says. 'He's always been eager to show off the latest developments to me. We haven't really hung out, as such. I think there's a bit of professional remove. He's pretty impressive. It would be cool to, like, actually, like, hang out with him, but I understand he's an extraordinarily busy man these days. It was weird because he had this period when he got tired, or burnt out, or he didn't expect RealDoll to explode to the extent that it did, and he had some sort of crisis where he was like, "I'm going to step away from the whole doll-making thing for a while," and he went to music.'

'When was this?'

'Good Lord, this would have been . . . I can find out, if you were to hang on for a couple of seconds?' He takes his headset off and goes to rummage for something off-camera. Sidore is still stuck in the frame, her purple hair fluttering in the wake of his movement.

Davecat comes back with a CD in his hand. 'He recorded two albums,' he says. 'This one is from 2006. It's pretty good stuff, actually.' He holds the CD up to his camera. The album is called *Hollow*. There's a photo of Matt posing between two band-mates, in full grunge pixie mode. Superimposed over it, in huge letters, is the name *NICK BLACK*.

'That was his pseudonym, Nick Black. That's him in the centre.'

I can't believe it.

'Like the *Nick* doll!' I say.

'Yes! That is his face. I guess at some point he realized he was doing a lot better as a doll maker than as a musician,' Davecat continues. 'At some point he saw iDollators such as myself having dolls as partners rather than just sex toys, and he realized if he could make dolls that have an artificial intelligence, he could make something huge. There's a bit of a renaissance thing going on with Matt. I think at this stage, he is content to improve the human condition through artificial beings.'

When I log off Skype, I get lost down a Google hole searching for Nick Black. I find a rarely updated Facebook page with 3,000 fans. One of the most recent posts is over a year old and says, 'Anyone who needs a copy of *Hollow* or *Awake* email me! I have a few boxes left!'

I find the Nick Black YouTube channel, which hasn't been updated much in ten years. There's a video for a power-chord-heavy song called 'Sorry', where Matt bounces and sings like Linkin Park's Chester Bennington and bites a model's neck with vampire teeth. Then there's an eleven-year-old seven-minute behind-the-scenes rockumentary that begins with Matt on a rooftop after dark. He looks out into the distance and says, 'Nick Black isn't just who I am, it isn't just the name of my band, it's an attitude. It's a way of becoming something more than you were.'

That didn't turn out to be true, of course: it is actually Harmony, rather than Nick, that has the real potential to make Matt something more than he ever was.

CHAPTER THREE

'It won't feel a thing'

Under humming halogen lights in downtown Las Vegas, Roberto Cardenas is making a plaster cast of a naked woman. He smears handfuls of gloopy pink casting gel all over her breasts and thighs, while his brother looks on and takes pictures. Softly spoken and awkward, with a nervous laugh and stiff, gelled hair, Roberto has the air of a mad professor, but he's as detached and clinical as a doctor taking a cast of a broken leg.

Matt told me he has no competitors: there might be a few Chinese companies trying to produce something made of cheaper materials that can move a little, he said, but those dolls are years behind the artificially intelligent girlfriends being made at Abyss. Yet the truth is there are entrepreneurs and engineers across Asia, Europe and the US who are racing him to put the first sex robot on the market. Just over the state border in Nevada, Roberto has spent four years working on Android Love Dolls, Eden Robotics' flagship creation, which he calls 'the first fully functional sex robot dolls ever made'. While Matt sculpts his idealized proxy females by hand, Roberto casts them from life, in a drive to make a humanoid so realistic it can't be distinguished from a real woman.

I'd found Roberto canvassing opinions from robot enthusiasts on dollforum.com. 'Hello. I am building an Android Sex Robot Doll and want to share my project with the community,' he wrote. He said his robot could 'perform +20 sexual acts', could 'stay upright by herself, sit down, crawl', could 'moan in pleasure during intercourse' and had 'speech AI for communication'.

'I am interested in knowing what features would the community like to see in a sex robot doll', he said. 'Thanks, and welcome to a new era in human–robot interaction'.

There were some links to his website, which showed a rather blank-faced humanoid in a suit jacket with sharp shoulder pads, and an unsettling video of a metallic robot skeleton writhing in the missionary position, which made me think of the final scene of the first Terminator film, after the cyborg's skin has been burnt away.

The replies came quickly.

'Eye contact would be nice', came the first.

'Voice recognition', came the second.

'Breathing is more important than the complexity of walking', said another.

'Make sure your gynoid has full body heat from head to toe', said a fourth.

The forum members were both sceptical and cautiously excited about Roberto's claims. 'There are many people on this forum that absolutely will buy one if you create a product we can accept', wrote one. 'We want you (or someone) to succeed'.

The men here didn't sound much like the disabled, lonely or socially excluded customers that Matt or Douglas like to talk about. Several mentioned their wives and girlfriends, and compared them unfavourably to their silicone doll mistresses.

One included a photo of his sex doll for Roberto to use as an aesthetic guide when planning his robot's proportions. She was in leopard print underwear, propped up in front of a wall adorned with daggers, mounted hunting knives and a bladed knuckleduster. 'If my RealDoll could cook, clean, and screw whenever I wanted, I'd never date again,' he wrote. 'That's what I really want, but that is just wishful thinking.'

*

I'd arranged to meet Roberto at ten a.m. at the artists' studio above a tattoo parlour where he works so we can have a chat before his model arrives. Las Vegas is a strange place at ten a.m. The tattoo parlour is padlocked and I can't find any other entrance to the building. I call Roberto and he tells me to go around to the back door, which is in an alley filled with discarded furniture and shopping trolleys. We have spoken on the phone and exchanged a few emails; he has sent me photos and videos of his robot that make it look like he's been working on something substantial. But I become very aware that I have absolutely no idea what I am walking into.

Roberto has thick glasses and a thick Cuban accent, and none of Matt's swagger; in every sense, he is at the other end of the scale to Matt. Eden Robotics is a part-time project for him; he makes his living as a pharmacy technician, measuring out pills behind a counter and never interacting with customers. Conversation doesn't come easily to him, but he's smiling broadly when we shake hands, pleased to have a journalist take an interest in the project he believes is going to make his name.

The studio is painted gloss black from floor to ceiling. Apart from a folding table, a white sink and a few boxes, it's completely empty – a dark, glossy void. Noel Aguila, Roberto's half-brother, is waiting for us, his arms folded across his Hawaiian shirt, in blue loafers and navy jeans. He's twenty-three, seven years younger than Roberto, and he left Cuba for the US six years before Roberto did, so he has a more American accent, and a more American kind of confidence.

'It's a new field in business, so we're learning as we go along,' Noel tells me as Roberto begins opening some of the cardboard boxes. 'I've been trying to help him with marketing and logo design, the website and exposure, trying to see the best way to sell it. Because the people who are involved with this are kind of . . . strange –' he grins – 'we've had some strange requests we've had to turn down. It's definitely different.' Noel, too, has a day job: he works at the box office of the Colosseum, taking tickets for Celine Dion and Elton John. He is used to facing customers, albeit those with more mainstream tastes.

Farrah, today's model, isn't here yet, but Roberto is busying himself while we wait for her, measuring out the casting gel, a pink powder called alginate, and mixing it with water in a white plastic tub. She will be the fourth or fifth woman Roberto has cast for Android Love Dolls, he says. This will be the first of many castings needed to make a complete mould of her entire body.

'What were you looking for when you found Farrah?' I ask.

'She's curvy,' Roberto says, only looking up briefly from his alchemy. He's had an order from someone who wants a fuller figure than the women he has already cast, so he is making one to the customer's specifications, but his market research has shown it would make commercial sense to have a larger model on general sale, he says. 'In the doll community, they are really interested in curvy girls with big butts.'

Farrah breezes through the door, a breath of fresh air. She wears a long-sleeved ash-grey ribbed dress that's skin tight and polo-necked and too warm for Vegas, hair pulled up in a messy bun, towering stripper heels. Her smile is dazzling and magnetic, and I'm grateful she's here; all of a sudden Roberto's awkwardness doesn't seem so contagious.

'Nice to meet you!' she beams. 'Who was I texting?' She looks at me. 'Was I texting you?'

'I'm a journalist,' I say.

'Nice to meet you!'

Roberto steps forward to shake her hand.

'So what exactly do you do the sculptures for?' Farrah asks.

'It's for an android robot,' he says. 'They're like dolls. They go into positions and—'

'So they are like sex dolls?'

'The first ones will be like that. Then later they'll be able to help you in the house. Like a housekeeper.'

'Interesting!'

Farrah found the job on Craigslist: $200 for two hours of being cast in plaster, and a $500 commission on every product sold with her body. 'I thought it sounded like a great job,' she declares. 'There's nothing else to do in the daytime in Vegas, except gamble. I hope mine sells.' She shoots Roberto a dazzling smile. 'She better be hot, or I'll be pissed off!'

We perch on the table while Roberto protects the floor by taping sheets of plastic across it. Farrah tells me she's been dancing and webcamming for eight years, and works nights at Spearmint Rhino to put herself through real estate school and support her seven-year-old son. Her parents are Iraqi and they don't know what she does for a living. I'm surprised to hear she's twenty-seven. She has the kind of voluptuousness that only very young women have: soft and curvaceous, without a single roll of fat.

'I was kind of sceptical when I first saw the job advertised,' she tells me quietly, while Roberto busies himself on the other side of the room.

'It sounded too good to be true?'

'Yeah, like I'm not going to get paid. Craigslist is kind of scary.'

Roberto shows Farrah how to stand – legs apart, arms away from her sides, palms facing forward, splayed just like those headless RealDoll bodies – and she peels down her dress to reveal nothing but a few tattoos: no underwear, no body hair. I tell her she should take her six-inch platforms off – she's going to be standing there for a long time and they make me wince just to look at them. Roberto begins to apply the alginate, starting with her shoulders. She smiles, uncomfortably. 'It feels like very cold toothpaste,' she says.

'Do you know what's going to happen to the cast of your body?' I ask.

'They had something similar at this year's AVN. They said that this is a new phenomenon, and it's going to be *big* – a robot who can interact with you and talk with you. I think it's fascinating that people can actually do this, that people will spend money on these things. Anything I can do to help. It's *cool*. Why not? Why not be part of the future?'

'But have you thought about the guys who will buy your body and what they'll do with it?' I ask, as Roberto applies generous swirls of gloop around her nipples.

'It doesn't bother me,' she replies, breezily. 'I think it's better than what I do when I dance, because those guys actually have me. When these guys have a bot, I won't be there.'

'You are literally being turned into a sex object here,' I say.

'Now that you put it that way, I'm sure that it's going to cross my mind, but it doesn't bother me. If anything, I'm helping someone with their intimacy. I think men have needs. Whatever they do, as long as I'm not there, I'm fine with it. Hopefully she's a big seller. That would be awesome.'

Farrah asks if she needs to spread her legs so that her

'actual vagina' is moulded, but Roberto tells her that won't be necessary.

'He's very calm for his job,' she says. 'He doesn't show much emotion.'

'He's a robot engineer,' I shrug.

'Right! That's true.'

Roberto takes care around the creases of her knees to ensure that every detail will be captured. Noel snaps more pictures. After the plaster-soaked bandages have been applied it becomes uncomfortable for Farrah: the cast is heavy, and the weight of it pulls on her body. She's hungry. But they have to wait for it to dry entirely before she can be freed from it, so Roberto tries to entertain her by taking out his phone and showing her a picture of his current prototype, Eva.

'Oh my *God*,' Farrah says. 'That's amazing. That looks so real. But her eyes are a little scary.'

'I need to put her eyeballs in,' Roberto says.

After ninety minutes, Noel and Roberto help Farrah peel herself out of the cast. They leave it face down on the floor, like an inverted, decapitated corpse. Every line on her skin, the folds of her belly button, every detail is there, in plaster, waiting to be copied in fiberglass and then reproduced in silicone. Roberto pays Farrah her $200 in cash and they make a plan for her to return, so he can cast the other side of her body, her arms, and finally her face. Everyone looks happy, no one more than Roberto. 'When I do something, I like to do it the best,' he beams. 'I want that level of detail. I want it so that you can't tell the difference between the robot and a real woman.'

*

Roberto knows I've come to Las Vegas to meet his robot, but Eva the Android Love Doll isn't at the studio today: she's in his workshop, which is the garage of the home he shares with Noel and their mother, in a gated community in the suburbs, twenty minutes' drive away. He brushes away the dog hair and plaster body parts to make room for me on the back seat of his car. Then he tells me how the robot has taken over his life.

'I have breakfast and a bath and then I work on the robot from eight until one. I go to work at the pharmacy until seven and then I come back and do a little more on the robot or the website. Right now, I'm working on the skeleton. For most of last week I was putting new, more powerful motors in the legs; the old ones were too weak. I work on it every day.'

Roberto is only in the US because his mother literally won the right to be here. In the 1990s, Cubans who qualified for refugee status could enter a lottery and win US citizenship for themselves and their families. She came over with Noel in 2000, while Roberto stayed in Cuba to take care of their grandmother, joining them after she died in 2006. 'In Cuba, people are hungry for technology,' he says. 'That's why I want to use technology to help people's lives.' He arrived in the US fuelled by American dreams of becoming a self-made entrepreneur, a rags-to-riches success. When he read a 2016 *Fortune* magazine article that predicted spending on robotics would hit $135.4 billion by 2019, he knew he had found his calling. 'I've always been interested in robotics. This is my passion. I love this. I love my job.'

He tells me his goal is to make fully functional humanoids that can model clothes and work the tills in the retail industry, show hotel guests to their rooms in the hospitality industry, and do domestic chores and look after the sick and elderly in the care industry. He's starting with sex robots because they are simpler. 'The movements are easier to do. A fully functional android robot would take a couple of years to finish – a sex robot is accessible now. It's the fastest way to achieve my goal.'

The whole family has bought into his dream: there's Noel, of course, in marketing and communications, plus their uncle, who helps Roberto in the workshop at weekends, but there's also their cousin, who's a year away from completing a PhD in cybernetics and is helping with some of the engineering. Roberto gets everything else he needs from Google, or YouTube, or Amazon. 'Mostly, I self-learn. I read books. It keeps me really busy.' The family has so far invested $20,000 of their savings into Roberto's prototypes.

'We're going to make it so her eyes can follow you. People in the doll community want warm skin, so I'm going to try and invent some sensors in the skin to raise the temperature – silicone can burn really fast, so I'm trying to see how I can do that safely. Some people have also said they want the doll to self-lubricate – I'm working on that. We are also interested in incorporating virtual reality technologies, so couples in long distance relationships can control the doll with their movements. We want her to have real relationships with people.'

Roberto sounds far more interested in developing the physical side of his robot than the companionship side of things. The AI – the possibility of having a relationship – is something he'll get to once he's cracked the animatronics. He tells me his ultimate goal is to build a robot that will walk up and knock on his customer's door. 'Self-delivery.'

Of course, Roberto has heard rumours about the work going on in Abyss Creations' RealBotix room, and the sex doll manufacturers in East Asia who are experimenting with animatronics. But he hopes that if he can beat them all and be the first to produce a sex robot that can put herself into sexual positions, he'll have the commercial edge. 'For full body movement, I'm pretty much one of the first ones,' he says. He's also undercutting his rivals on price: his robots will cost $8,000 to

$10,000, and five customers have already paid for theirs in advance.

By the time we drive into the compound and pull up outside Roberto's garage, Eva has had quite a build-up. He flicks the switch to pull up the garage door to reveal his workshop, and it feels like a curtain is being very slowly lifted.

Eva, the robot he claims can put herself into over twenty different sex positions, the robot he says can crawl and moan and has fully functional AI, the robot he told me was 'ready twenty-four seven', is lying headless and footless on a trestle table at the back of the garage. Her metal skeleton is clearly visible under her silicone skin, which has thick, jagged seams. It looks like a mess.

'Let me just get the head,' Roberto says, shuffling inside the house, with Noel a few paces behind.

The workshop is a monument to Roberto's obsession. Another headless silicone body reclines on a mattress in the corner. The yard next to it is filled with shop mannequins, torsos, a pair of legs with purple painted toenails, and a large cardboard box filled with plaster casts of human heads. The garage floor is carpeted with Newport cigarette butts smoked down to the filter.

The brothers re-emerge from the house with the blank-faced head in a brown wig that I recognize from the website, some itchy-looking thick black stockings and crotchless white panties bedecked with pink bows. Roberto dresses Eva with fumbling hands, screws the head onto the neck and plugs it into a laptop resting on a battered leather chair. But Eva is not going to perform for me today. Roberto tinkers, reboots and rewires, but her sound files won't load, he says, and her new limbs are too heavy for the existing servomotors, so she can barely move. Her joints wheeze as he tries to get her to bend her legs.

'It's all trial and error, at this stage,' he shrugs, without any embarrassment at all. 'She's a prototype.'

Roberto has complete faith that his robot will exist one day. He is determined to make his dream a reality, to prove to his family that their belief in him, that their investment in him, is justified.

'Do you have any worries about making a robot like this?' I ask him.

'No, not really. It's a technology that's moving forward, and pretty soon robotics and technology will be more and more in our daily lives. It will help people to become more sociable.'

'So it's perfectly healthy to want to own a robot you have sex with?'

Noel the marketing man senses a change in tone and steps in.

'Women experience things like rape and abuse and things like that,' he says solemnly. 'This is definitely something that could help people move away from that, so they are not so angry with their wives: they can be angry at this, and beat this, and that should be fine –' he throws open his arms – 'because it will not feel a thing, we promise!'

The brothers laugh, open-mouthed, delighted with the joke. But Noel isn't really joking.

'Hold on a second,' I say. 'Surely people like that should be encouraged not to have those feelings at all, rather than being given something to rape and beat.'

'Yeah,' nods Noel. 'These things will help them, calm them down, and act as a safeguard between anything that they want to do and anything they *will* do.'

I leave Roberto and Noel just as their mother, Marilyn, is arriving home from work. She wears a large crucifix on a thin chain around her neck. I'm desperate to know what she thinks about her son's project.

'I think there's a genius in my garage. Like the Apple person – Steve Jobs – I saw the movie,' she says warmly, her face flushed with joy. 'He has a great idea and he concentrates hard on the job. I told him that he can reach for the stars. The sky isn't far for him.'

'You seem so proud,' I say.

'He's very capable of achieving his goal. He's an intelligent boy.' She puts her hand to her heart. 'He's my son.'

I head back to my hotel as the reassuring cloak of darkness falls on Las Vegas. I'm exhausted. Music is thumping out of huge speakers mounted on the building's exterior: throbbing, pounding beats that are supposed to entice gamblers into the hotel's casino. I swipe my key card and flop down on the giant bed. On the bedside table, there's a metal dish full of individually wrapped pairs of earplugs: wax ones, foam ones, silicone ones – a profusion of solutions supplied by the management to the noise pollution problem caused by the management. They could just switch the music off, of course, but they have provided a little piece of technology instead so they don't have to.

My head is full of Eva, who has the body of a real woman, but can be beaten without feeling a thing. Rather than dealing with the cause of a problem, we invent something to try to cancel it out.

*

Sex robots are coming onto the market during a time of turmoil for men around the world, when they are losing their power, their status, their certainties. The sexual revolution and second-wave feminism of the 1960s have meant that today, in the West at least, women grow up knowing they can and should choose who they sleep with. They are no longer viewed as the property of fathers, passed down to husbands. They feel entitled to fulfilling relationships, and are less willing than ever to stick things out when they are not good.

Some men have found this reimagining of women as sentient beings with desires and choices very inconvenient; it has left them without access to sex, and it's made them very angry.

'Incels' are the self-proclaimed involuntarily celibate. Although some women have identified as incel, the term has been overwhelmingly adopted by heterosexual men, who believe they are entitled to sex with desirable women whenever they want it, and loathe women for denying it to them. They think women should be easier, at the same time as being disgusted by how easy they are. Their special brand of misogyny despises women for refusing to have sex with them, without considering that the reason women don't want sex with them is not because they are not rich or good looking enough, but because they are misogynists.

In their online message boards, incels say that women use their sexual power over men to tyrannize them. They describe themselves as a marginalized group fighting for their right to sex in the face of terrible injustice, just as black people are fighting for their right not to be killed by police officers. I've read posts where they lament how women are 'worshipped' when they are only 'cum-dumpsters', how they need to be murdered, stalked and 'raped in the eye sockets'. It would be easy to dismiss this as simply the online rantings of a few desperate losers, but there is a worryingly large number of them. When Reddit shut down its online incel community for glorifying rape and violence against women in November 2017, the incel subreddit had 40,000 members – that's *members*, people actively contributing to the message boards; it doesn't include the people who lurk and read the page without signing in – and it was only one of scores of similar communities online.

And incels aren't just hiding behind their computers: they are radicalizing one another and committing mass murder. At least sixteen people have been killed by men happy to describe themselves as incels. In 2014, in Isla Vista, California, Elliot Rodger killed six people and injured fourteen others before killing himself. Shortly before the attack, he uploaded a YouTube video in which he told the camera, 'I don't know why you girls aren't attracted to me, but I will punish you all for it.' Four years later, Alek Minassian drove a van into a crowd of people in Toronto, killing ten and injuring sixteen, just after posting on Facebook that 'The Incel Rebellion has already begun!' Many more have died at the hands of men who said they were motivated by sexual frustration – like the Virginia Tech shooter Seung-Hui Cho, who killed thirty-two people in 2007, and Christopher Harper-Mercer, who killed nine people in Oregon in 2015.

So sexually frustrated men can be dangerous. And it's not just Noel who thinks sex robots can be the solution; think pieces from the *New York Times* to the *Spectator* have suggested that, in the future, sex robots will be used to defuse and pacify the involuntarily celibate before they can do any harm to humans. The argument goes that sex robots will allow for a kind of 'sexual redistribution' which will mean the right to sex can become an attainable human right, and that life will no longer seem so terribly unfair for men who can't get laid.

But sex robots are more likely to be a symptom of the problem than a cure for it. They have been developed at the same time as both incel culture and deepfaked pornography, where faces (of celebrities, or ex-partners, or anyone, regardless of whether they consent) are superimposed into porn videos. It is not enough for porn to exist for free, in our pockets, whenever we want it; some men want the precise porn they'd like to see, even if their desired actors don't want to make it. Deepfakes allow anyone to be made into a pornographic spectacle without them knowing or feeling a thing.

To an even greater degree, sex robots can offer total control for the men who want it most, the chance to have a partner without autonomy, a partner they can dominate completely, stripped of the inconvenience of her own desires and free will. A partner who is built like a porn star, but will never gag, vomit or cry. For these men, this would be an upgrade on a real woman. Sex robots who never say no will feed this kind of desire, not extinguish it.

There are manufacturers in China and Japan who have no qualms about producing child sex dolls. They argue that giving men who are attracted to children a synthetic substitute will stop them from abusing real children. Men across Europe and North America have been arrested for trying to bring them into their countries (in the UK, our archaic laws mean that it's the importation of child sex dolls that is illegal, rather than the use of them). Whenever they make the news, there's almost universal disgust that child sex dolls could ever exist. A few tenacious academics have speculated about whether owning a child doll could stop paedophiles from acting on their impulses, as if they could be a kind of substitute for children in the way that methadone is for opiates. But the general consensus seems to be that there is no safe way for paedophiles to act on their urges – that, instead of sating their desires, child sex dolls would feed them.

None of the people racing to release the world's first sex robot is trying to market a child model, not even Douglas, whose 'Young Yoko' version of Roxxxy True Companion is so carefully described as 'barely' over eighteen. But if child sex dolls are taboo because they could encourage illegal, damaging and abusive behaviour, how would allowing men to act out their darkest fantasies on female robots be any different? If the existence of child dolls could harm real children, how can we be confident that female sex robots pose no danger for real women?

Of course, the 'manosphere' of extreme male rights communities online loves the idea of sex robots. We will hear plenty more from them when we look at the future of birth, but, for now, forgive me for reproducing a few comments in full from mgtow.com, the site for Men Going Their Own Way, complete with faux-virtuous censored obscenity and original punctuation and syntax:

Time to replace these c~~~s with robots !

The end of thousands of years of female c~~~ dictatorship

In the Book of Genesis God created woman and promised us her as a 'help-mate'. Someone to help us, obey us, someone warm, caring, supportive and empathetic . . . Well we didn't get it–did we? instead His creation has been corrupted to the point that it is anything but what it was suppose to be. (women) So, we shall make our own help-mate and then we will finally have the companion promised to us by God.

The comments on this particular thread were in response to a news article about Dr Sergi Santos, the Spanish engineer based in another garage workshop, six thousand miles away from Roberto's, in Rubi, just outside Barcelona, Spain. Sergi is the fourth person I have found to lay claim to having invented the world's first sex robot, but unlike Matt, Roberto or Douglas, his robot began life as an academic project, an experiment in machine learning which he documented in a paper for the *International Robotics & Automation Journal* entitled 'The Samantha Project: a Modular Architecture for Modeling Transitions in Human Emotions'. He has a PhD in nanosciences – the study of the properties of tiny particles – but has spent the last four years working on a model for an artificial theory of mind.

Sergi only planned to design a brain at first, but when he was looking for a credible body in which to house it so that humans would interact with it authentically, his wife, Maritsa Kissamitaki, stumbled upon the world of hyperrealistic sex dolls. Sergi spent $50,000 buying ten from around the world, including a RealDoll and several cheaper Chinese models, and turned them into robots, adding a microphone, speakers, an internal computer and touch sensors so the doll could respond to human touch and learn from human interaction. He called her Samantha because the name means 'the listener' in Aramaic.

Maritsa worked out how to put the sensors in the body; originally a graphic designer, she turned herself into an expert in robot assembly. Samantha doesn't have much in the way of movement – her vagina vibrates and there's a motor in her jaw; she moans and talks, but her lips don't move – yet this means the Samantha system could, in principle, be used to bring any sex doll to life, and could be sold for far less than even Roberto is asking. By focusing on the computational side – the software, rather than the hardware – Sergi was potentially making sex robot technology available to a far wider range of people. His company, Synthea Amatus, claims to have begun selling it in 2017, with a starting price of €2,000.

There are several ways of running Samantha, ranging from 'hard sex' to 'family mode'. She makes a lot of noise when she 'climaxes', and, by responding to the sound and movement of her owner, she can learn how to fake a simultaneous orgasm. 'Samantha will call you and ask you for attention,' says the Synthea Amatus website. 'The more she has to ask for attention, the more patient she will become, the more you pay attention, the less patient she will become. She will learn to not call continuously.' This version of ideal femininity will yawn and go to sleep if you ignore her, but will never be too tired for sex. 'If you interact with her in this relaxed state, she might get sexually excited. If you leave her she might cool down again and fall back to sleep.'

When Sergi first went public with news of his creation, he was happy to talk to everyone. Some of his interviews would best be described as ill-advised. 'I'm basically the Robin Hood of sex because I give to the poor. Men need sex and I just give it to them,' he told a reporter from ITV, his arm draped across Samantha's shoulders. 'Women and men view sex in a very different way. Men want more sex. A man wants to feel in general that the woman is desperate to have sex with him.'

I think everyone wants to feel desperately wanted when they are having sex, but women probably find it harder to suspend the disbelief that a substitute human made of silicone actually desires them. Sergi isn't thinking about female desire, though. His view of sex is self-centred, to say the least.

Reporters latched on to the detail that Maritsa, Sergi's partner of sixteen years, was working alongside him. The couple gave joint interviews about how their marriage had been enhanced by Sergi's private use of Samantha. 'I need sex some times of the day when my wife doesn't want to,' he revealed on the YouTube channel Barcroft TV, while Maritsa lingered demurely to the right of the shot. 'I could have sex three or four times a day,' he told a BBC crew, whom Maritsa told quietly in a separate interview, 'It can calm him down. He has a bigger drive than I do. If he's calmer, then it makes the day easier for all of us.'

Sergi was quoted talking as if men's insatiable libido was taken as given, as if sex was something men needed and women often had to either deny or endure, and as if he had invented a machine to help both men and women by removing the problem of 'lack of synchronicity' in a couple's sex life. There was no mention of Sergi's theory of consciousness and how the robot was an academic project to understand the human brain by modelling transitions in human emotions. The coverage was all about a sex-mad scientist and his long-suffering wife.

By the time I speak to him, Sergi has fallen out of love with journalists. We have some long conversations over Skype, but he tells me he doesn't want to give any more interviews, and certainly doesn't want to put anyone in contact with his wife after the BBC filmed her. 'How did I allow these guys to talk to my wife in a room on their own?' he says, oblivious to how much that makes him sound even more like a caveman. 'Unfortunately, I don't want anything to do with the media now.'

Besides, he's quitting the sex robot anyway. 'I didn't do this for money. I try to learn, to see what it is, and build it,' he says. He's handed the business over to his manufacturer, and if there's a demand it will be met, but he no longer wants to be involved in developing the product further. His experience in trying to bring the robot to market has made him lose faith in humanity. 'There's more humanity in this doll here than in the journalists I've met,' he tells me, pointing at something silicone in the corner of his workshop. 'And actually, for me, the doll is a way to become more human.'

But for the online army of misogynists who respond so positively to every new headline about the coming of the sex robots, Samantha and Harmony and Eva and Roxxxy are attractive because of their very lack of humanity; they are desirable because they can't think and feel and choose for themselves. Sergi might have started work on his robot in order to better understand the human brain, but he's ended up at the beginning of a production line that has the potential to erode our empathy – the beginning of the end of human relationships.

CHAPTER FOUR

'All of our relationships are at stake'

The *Robots* exhibition at London's Science Museum is like the greatest hits of robotic engineering, a roll call of the world's best-loved humanoids. There's Harry, Toyota's Partner Robot, who sways and bops as he plays a jaunty tune on the trumpet. There's the Honda P2, the first robot to walk like a human, whose bubble helmet head and cream body make him look just like he's wearing one of the astronaut suits on display in the space gallery downstairs. There's Pepper, the cute little companion robot with anime eyes, who fist-bumps the delighted visitors who are queuing up to meet him.

'What we are seeing here is the graveyard of the modern individual –' Dr Kathleen Richardson frowns – 'the idea that we're just machines.'

Kathleen has not come here to fist-bump Pepper. She is the director of the Campaign Against Sex Robots (CASR), founded in 2015 and launched at an ethics conference at Leicester's De Montfort University, where she is Professor of Ethics and Culture of Robots and AI. I'd arranged to meet her at the exhibition thinking it would be a colourful place to hear about her campaign, but even though the robots here are distinctly non-sexual, Kathleen does not see the fun in them.

'Starting the campaign feels like a very necessary response to what I consider to be a very dark period in humanity's progress,' she tells me, as the robots hiss and whir around us. 'We live in a world that tries to convince all of us that we are not connected to each other as human beings, that actually we are alone in the universe, we're born alone and we die alone, and we can use other people as our forms of property. This exhibition is a tribute to modern individualism, a society that now wants to interact with objects as though they were like other human beings.'

The campaign is 'a group of activists, writers and academics developing new, and sorely needed feminist and abolitionist perspectives of robots and AI,' according to its website. They are calling on government ministers to legislate against the rise of the sex robots 'before it's too late.'

'We believe the development of sex robots further sexually objectifies women and children,' says their mission statement. 'We take issue with those arguments that propose that sex robots could help reduce sexual exploitation and violence towards prostituted persons, pointing to all the evidence that shows how technology and the sex trade coexist and reinforce each other creating more demand for human bodies.'

There's a large, unsettling black and white photo of Kathleen on the website, taken in front of a wall covered with a collage of nightmarish images of Maria, the iconic humanoid from *Metropolis*. Kathleen is dressed in black, with a black, messy fringed bob and no make up, her dark, unsmiling eyes staring directly into the camera lens. By her very non-conformity, she conforms to the stereotype of what the online manosphere might imagine an angry feminist to look like, and she makes no apology for it.

'Sex dolls rest on an idea that's already present in society: that women are property, that women are not fully human, they are subhuman, and they can be related to as a form of property,' she tells me, while Kodomoroid, the uncomfortably realistic

Japanese gynoid newsreader, bows respectfully behind her. 'Creating a robot that you can now have sex with is a logical consequence of the idea of the modern individual as separate, atomized and disconnected from others. Sex is an experience of human beings, not bodies as property, not separated minds, not objects. It's a way for us to enter into our humanity with another human being.'

Kathleen's take is as much Marxist as feminist: she thinks sex robots are a symptom of a consumerist society gone to excess; they embody the worst elements of unbridled capitalism because they turn relationships into commodities. 'The people who make them are saying they're not just a masturbatory tool, they're taking the logical idea of the individual to its extreme, they're saying, "You can have a relationship with this doll. It can be your girlfriend. It can be your wife. In the future, you'll be able to marry these dolls." This continuous isolating negative force is acting on our relationships.'

There's a lot to unpack here. 'So sex robots threaten human interaction?' I ask.

'Absolutely,' she nods. 'In fact, human interaction is already threatened by the rise of technology today because it rests on the idea of the individual. Think about it: the *i*Phone, the *i*Pad. It's all about the "I".'

I do think about it, and I'm not sure I follow, but Kathleen is in full flow.

'People in power don't want people to get together and make relationships with each other; they want to turn them into isolated, individuated atoms that consume products. An Oxfam report came out today saying eight human beings currently own half the world's wealth. As human beings who are not members of those elites, the only thing we have is each other. If we take those steps to abolish those practices that keep us isolated and separated from each other, we might stand a chance to do something different in our world.'

'Is the answer banning robots?' I ask.

For the first time, Kathleen hesitates. 'Museums are good places for robots. Certainly, we should have automation in our lives – that can be quite helpful to us, as human beings. But the problem once again comes from the concentration of power in the hands of a few.'

In fact, the CASR doesn't have a clear position on whether they are calling for sex robots to be outlawed. At first they demanded a ban, then they called for a serious examination of the ethical consequences, and then they campaigned for 'public consultations ahead of developing legislation', without specifying what that legislation should be. Kathleen's campaign is more of a critique than a movement, and it's not a neat, easy-to-grasp critique, either: it relies on very specific, academic definitions of personhood and sex – a premise within a very particular worldview. One that's a world away from Farrah, or Matt, or Davecat.

'I've met some of the people who are making these robots. They say they are just trying to make people happy. They say their robots are therapeutic, they are about creating an illusion of companionship for people who otherwise would not have a chance to have any companionship,' I say.

'It's a myth. And it's a lie, actually,' Kathleen replies. 'Every human being has relationships. We are not isolated.'

'What about having someone to come home to? Someone they can talk to when they might otherwise never have a conversation?'

'If you have these objects in your life you are still alone. People and objects are not interchangeable.'

'So they are being kept alone?'

'Yes. And the objects then start to fill the places of other human beings, of hurt feelings, of suffering, of despair, of loneliness,' she continues. 'I would call that part of rape culture. The more that they participate in activities that are outside of this consensual framework, they then turn themselves into objects.'

Kathleen may express it in an uncompromising and sometimes impenetrable way, but she has a point. Objectification isn't only about encouraging people to look at human bodies as if they are things, like you might gawp at the pornified breasts and impossible waists in the Abyss Creations workroom, it's also about treating people like objects. The global trade in human bodies for sex work – human trafficking – is a booming industry that depends on viewing women and children as nothing more than cargo, to be transported and used, like drugs or arms. Any product that encourages us to see people and objects as interchangeable also feeds into the mindset that enables slavery.

'This is not stopping,' Kathleen says. 'This is a train that is going quickly and it's going at a speed that no one really understands.'

We take a walk around the exhibition, past ASIMO the dancing robot, RoboThespian the performing robot, and Zeno, the robot boy whose expressive face will mirror any look of anger, happiness or surprise it detects on yours. Signs dotted around the hall are supposed to make us think deep thoughts. *Is it ethical for a robot to pretend to be human?* they ask. *Would you be friends with a robot?*

'Would you be friends with a robot, Kathleen?' I ask.

'It's impossible to be friends with a robot because our experience of friendship comes from human relationships. These are inanimate objects.'

She sounds almost robotic.

The Campaign Against Sex Robots got a lot of press when it first launched, but mainly because journalists liked the idea of the campaign, rather than the campaign's ideas. It was an excuse to tell the irresistible story of the dangerous, perfect man-made partner that has always so beguiled us. Journalists weren't interested in probing into whether the feminist abolitionist approach to property relations was the right prism through which to view sex robots; they just wanted any excuse to view sex robots. It was ironic that, given the campaign is a reaction to uncritical reporting of sex dolls and sex robots, the person reporters called on to give the counterargument on behalf of the sex tech industry was Douglas Hines, the man who probably doesn't have a robot to sell, after all. It didn't matter, so long as the story was good.

But Kathleen doesn't care about a good story. She tells it as she sees it, even if she sees it in a way that is going to alienate a lot of people. The first time I ever heard her speak she was giving a lecture at the British Academy in London, and the room was packed, with rows of people standing at the back.

'I'm thinking about changing the campaign to the Campaign Against Rape Robots, because that's actually the most appropriate name for them,' she told the crowd. 'The moment sex stops being simultaneous is the moment it becomes rape.' She went further. 'In prostitution, women are raped. It's paid rape. In pornography, the performers are prostitutes because they are paid for sex. Pornography simulates the experience of rape for the viewer. If you watch pornography, you are imitating a rape fantasy.'

This was too much for the audience of feminist millennials who had grown up in the age of ubiquitous free porn and would never consider themselves rape enablers. Some of them were openly laughing at her new set of definitions.

'The world of sex robots is mimicking this cruel form of rape that's now become mainstream and normalized in our culture. That is a problem for every single one of us. All of our relationships are at stake here,' she implored the audience. But she had lost a lot of them.

While Matt and Roberto tinker in their workshops, fundamental questions need to be asked about the implications of what they are doing. But Kathleen might not be the person to ask them.

*

It's the second International Congress on Love and Sex with Robots, and the 250-seat auditorium of Goldsmiths' Professor Stuart Hall Building is packed. Academic delegates sit in the middle of the room: geeky men and women in their twenties and thirties with avant-garde haircuts: super-short fringes, experimental sideburns. On the left of the auditorium, near the exit, perch reporters who have flown in from across the globe to file breathless copy about any new developments in the world of sex robots. Most will leave disappointed; this is a series of academic lectures about humanoid robotics, not a demonstration of the latest hardware.

Computer scientist Dr Kate Devlin is bouncing with excitement as she takes to the podium to give her keynote speech; people in her field aren't used to journalists being so interested in their work, she jokes. The second International Congress of Love and Sex with Robots was supposed to be held in Malaysia, but the Muslim country's Inspector General of Police banned it only days before the event, on the grounds that it was promoting 'an unnatural culture'. It has made the conference notorious. 'This isn't a sex festival,' Devlin tells the journalists. 'We're thinking about some really big issues.'

Co-founded by David Levy and named after his book, the two-day event is in many ways an attempt by the academics who see potential benefits in human–robot relationships to address the criticisms raised by Kathleen. She hasn't been invited to speak today, but her arguments hang heavy in the air, and many of the speakers use their time on stage to respond to her. Devlin argues that instead of campaigning against sex robots, we should use them as an opportunity to explore new kinds of companionship and sexuality. It's something she has covered extensively: as well as being one of the few computer scientists who specializes in sex tech, she has written about her own polyamorous relationships, and how 'consensual non-monogamy' has enriched her life.

If current conceptions of sex robots objectify women, Devlin says, we should work to reshape those ideas, not try to repress them. 'It can go somewhere else. Why does a sex robot have to look like a person?' she asks. Advances in smart fabrics and e-textiles mean we could make abstract, immersive sex robots that can envelop and embrace you, cuddly sex robots upholstered in velvet or silk, robots with 'mixed genitalia; tentacles instead of arms', she says: our attraction to the humanoid form is just a habit. I try to imagine if a horny robotic teddy with tentacles could ever have mass appeal. I can't see it happening. Millions of years of the evolution of desire mean we are turned on by the human shape. Otherwise wouldn't we be humping branches, or bushes, or pebbles? It will take more than clever textiles to rewire us.

Then she talks about Paro, a fluffy, white, AI-enhanced robotic seal pup from Japan, who squeals and blinks its long eyelashes, and is charged by inserting a plug designed to look like a baby's dummy into its mouth. Paro has been used as a therapeutic pet for people with dementia across the world, from the US to Germany to NHS care homes in the UK. 'Paro doesn't need feeding, doesn't shit all over the carpet, no one wants to have sex with Paro,' Devlin jokes. Companion robots like Paro have brought great comfort to people who would otherwise have little contact, and sex robots could take that fulfilment a step further, she argues. There is something so terribly sad about people in nursing homes having robotic pets when what they really need is human contact, but the assumption seems to be that robots are more dependable than humans. 'To ban or stop this development would be short-sighted, as the therapeutic potential is very good,' she says. 'It's not necessarily going to be a terrible thing.'

Devlin says other issues posed by sex robots are far more pressing: they could easily betray you by divulging your data. And smart sex toys have already done this: in March 2017, the Canadian makers of the We-Vibe vibrator paid out a $3.75 million settlement in a class action lawsuit after it was revealed its makers were collecting real time data on how often its 300,000 owners used the vibrator, and at what intensity. Later that year, Hong Kong-based smart sextoy maker Lovense's remote control vibrator app was found to be recording the sound of some users' masturbation sessions without their knowledge and secretly storing the audio files. Once a robot like Harmony is on the market, she will know a lot more about her owner than a simple vibrator ever could. What if this information fell into the wrong hands?

Problems with smart sex toys have also revealed the potential for sexual assault at the hands of robots. The American Siime Eye vibrator, which has a built in camera so users can 'record and share' their sessions, has been found to be easily hackable, meaning that the incredibly intimate videos it takes could be stolen, but also that control of the device could be hijacked by strangers. Devlin doesn't mention it, but Lovense's Hush butt plug was discovered to have security problems that meant it could be remotely controlled by anyone within Bluetooth range. Hacked sex robots have the potential to cause scenarios even more nightmarish than butt plugs gone rogue.

My mind reels when I think about how lucrative it could be to sell the data sex robots collect from their owners to advertisers. Matt's words flood back into my head: *She systematically tries to find out more about you, until she knows all the things that make you you, until all those empty spots are filled.* Forget Cambridge Analytica and Facebook – there is a future where the information your partner has learned from you could be sold to the highest bidder. And then the being you love and trust the most might be used as the most powerful marketing tool ever known, making suggestions and recommendations to try to convince you to buy stuff. Or vote for stuff. Sex robots could entertain you, satisfy you, but also humiliate, hurt and exploit you. Perhaps there is no such thing as the perfect, true companion after all, human or humanoid.

Levy takes to the stage to thank Devlin. 'I'm glad that someone has the courage to speak out against Kathleen Richardson,' he says. 'Would you be against the idea that a sex robot would keep data about the experiences that it has had with its human lover in an attempt to become a better lover and for its human partner to become a better lover? Sex robots could use learning to great advantage.' As ever, Levy is determined to see the positives.

Sex robots appear to be the perfect blank canvas onto which to project your own personal beliefs and hang-ups, even if you would never consider having sex with one of them. If you are a male libertarian computer scientist, they are a brave new world of opportunity. If you are polyamorous sex tech specialist, they offer a way of exploring unconventional kinds of sexuality beyond what Devlin calls the 'monoheteronormative' mainstream. If you are a Marxist feminist, they represent the commodification of women. The debate currently taking place about sex robots reveals more about us today, our current desires and fears, than it does about the future of sex.

At the close of the day, Levy makes a casual remark that lingers with me. No matter what Kathleen might be campaigning for, he says, there is nothing anyone can do to stop the rise of the sex robots. 'I don't think anything to do with ethics or morals is going to stand in the way,' he continues. 'I really don't think it's possible to stop the world from developing something the world wants to develop. There are too many countries, too many rogue states, too many commercial interests.'

And of course, he's right. While academics tie themselves in ethical knots in the UK, the Chinese have been quietly getting on with the job.

*

Two of the most enduring clichés about East Asia are that, first, it's where technology advances unconstrained by any ethical boundaries and, second, it's the home of the world's weirdest attitudes to sex. People in China, Korea and Japan are supposed to be both sex obsessed and sexless, an incoherent and unfair stereotype, especially given that much of the demand for the strangest sex toys made in this part of the world comes from North America and Europe.

But it's true to say that East Asia is where most of the world's sex dolls are manufactured, and it's also where the most startlingly realistic humanoid robots are developed. Take Hong Kong-based Hanson Robotics' Sophia, the robot with fifty different facial expressions and the first humanoid to be made a full citizen of a nation (the Kingdom of Saudi Arabia, a country that doesn't grant citizenship to human refugees and is probably not the best place for any female, synthetic or organic). Or Geminoid, the famously uncanny robot the Japanese engineer Hiroshi Ishiguro created in 2007 to be his identical twin. As Ishiguro gets older, he keeps the same hairstyle and has regular corrective plastic surgery so he continues to look exactly like his android doppelgänger, an effort that is both vain and in vain.

Matt may have tried to give me the impression that I'd be wasting my time if I looked into the sex robots from East Asia, but he knows very well that this is where the greatest advances in humanoid technology are being made. And his biggest competitor is here, watching his every move from a port in a peninsula that juts out into the Yellow Sea.

If Abyss Creations is the Apple of sex dolls, Doll Sweet is Samsung. Based in Dalian, one of China's busiest seaports, DS has been making and shipping their DS Doll line of sex dolls since 2010. They sell around 3,000 a year, mainly to Japan, Europe and the US (Miss Winter, the one doll in Davecat's collection that he doesn't 'get romantic' with, is a DS Doll). Like RealDolls, DS Dolls are ultra-realistic, handmade in a custom blend of silicone, fully poseable and customizable, with faces cast from clay sculptures, and feet and hands cast from life. But they are cheaper and faster than RealDolls: you can buy a full doll for $3,000, and it will only take them a week or so to make it.

And DS Dolls are beautiful. They have delicate, perfect features and none of the brassiness or pornified proportions of their American competitors. Some of the dolls have very young-looking faces (although always on adult bodies), but *Fleur* and *Serena* are clearly supposed to be mature models, with crow's feet and dark circles under their eyes. (There are no sagging breasts or middle-age spread in their selection of bodies yet, though.) Most of the face options are Asian, but there are European ones too. 'We create beauties and dreams,' says their English language website. 'Our mission is to promote openness, innovation, towards progressive and more perfect development.' In this spirit of openness, the site has an unintentionally hilarious video of a man in a lab coat and white gloves, whose face is never shown, clinically palpating a doll's breasts to demonstrate their lifelike bounciness, while an instrumental piano version of Abba's 'Dancing Queen' tinkles in the background.

DS Robotics was launched in 2016, several years after Matt started work on Harmony, but DS spent far more, far quicker than Abyss: $2 million in the first two years of research and development. The videos they have released of their prototype make Harmony look prehistoric. Their robot has a fully expressive face: she can wink, raise her eyebrows, grimace and guffaw, and her smile is warm and believable, with no hint of Harmony's eyeless sarcastic smirk. Her arms and upper body move, and she tilts her head lyrically when she talks – or sings, as she does in Chinese in one of the videos, closing her eyes and swaying, as if lost in the music. DS have focused on getting the animatronics right, and the AI is an afterthought so far, little more than you would find in Siri or Alexa, meaning their prototype looks and feels incredibly real, but doesn't sound it. Yet.

It took four months of emailing, but I finally have a video call scheduled with Steven Zhang in Dalian. He's the chief development officer of DS Robotics and he appears in some of the videos goofing around with the prototype. In one, she screams and startles him, making him spill water all over his white lab coat; in another, he squirts breath freshener into his mouth and gives her a peck on the cheek that makes her roll her eyes and dry heave. His background is in movie effects, special make up and 3D animation, so he is used to making his creations perform.

When I finally see Steven, he is serious, professional, with the presence, confidence and authority of a man who leads a team with a multimillion-dollar budget. The white coat is off; he wears a blue shirt buttoned up to the neck and glasses with thin, tortoiseshell frames. The lab around him is bright and busy – there are thirty people in the robotics department, and many of them are working together at a huge pine table, next to a wall filled with racks of electronics.

'There will be a very huge market for the robot, and we want to get into that market,' he says, in almost perfect but thickly accented English. 'Very huge, I think, not only in China. In future, many people will need robots to do a lot of work that can help people.'

'You mean service robots?' I ask.

'Yeah, like in the government, in the office, in restaurants and movie theatres. Anywhere you would see people – like waiters, serving people – you will find robots.'

'Then why focus on robots you can have sex with?'

'The sex robot is just a small part of the function,' he says, with a gentle smile; it reminds me of Matt's exasperation that I kept focusing on the sex when he had made a robot that could do so much more. 'Maybe some people want the beautiful, the hot-woman type robot with the sex function, but that's not the main point.'

The main challenge DS Robotics is grappling with is the uncanny valley, which makes sex robots unsexy, Steven says. 'We have been in the adult product market for many years. We know when people want silicone dolls they have some beautiful image in their head. When the sex doll is sitting in the chair or lying on the bed, they can still have that image. But when the sex doll starts to do some actions, that totally breaks the image.' At the moment, sex robots aren't convincing enough for owners to suspend their disbelief, but they have the potential to shatter the imaginary world doll owners create around their dolls. 'At this stage, the technology can't replace real humans.'

'One day the technology will be good enough, though, won't it?'

'Yes. We hope that day comes soon,' he replies, with another soft smile.

He takes me on a Skype tour of the lab. Men with bowl haircuts are hunched over LCD screens. The two prototypes I recognize from the videos are at the far end, near the window. There is a delicate, elegant robot with long, tousled hair, dressed in a pastel blue cheongsam adorned with embroidered flowers, who bows demurely and says, '*Nǐ hǎo.*'

'We hope this one could go in the front door of some store,' Steven says.

The other robot has exposed circuitry at the back of her head. There's skin only on her face, neck and shoulders; the rest of her is a dark, intricate skeleton with a full set of ribs. Steven picks up a robotic arm covered with pale flesh and brings it back over to his desk to show me how it moves. It's bewildering how graceful steel and silicone have become in this laboratory.

'How close are we to a full body robot?' I ask.

'Right now, the arms and the top half of the body move, and the face. I think maybe next year.'

'Will it be able to walk next year?'

He nods emphatically. 'We try to do that.' He does a little walking action with his fingers, making them scuttle across his desk. 'We hope that people won't be able to separate real humans from robots in future, so that relations between humans and robots will be better.'

'In what way will it make it better?'

'In many ways. Let me think how to say this in English. Right now we can buy robots from eBay or anywhere else that can help people clean their rooms. There are also robots that can cook. We can already buy them very cheaply. But they don't look human. When people have choice, they would like a beautiful girl or handsome man to help them clean their room and cook, not some movable trash can.'

'So your idea is that in the future we will have service robots that can do everything for us? They can cook, they can clean, and if we want to have a relationship with them, we can?'

'Yes,' Steven nods with enthusiasm. 'That's right. In future.'

'You are using what you've learned at DS about how to make very realistic dolls that look and feel human, and you're adding that technology so people can have a service robot at home that they treat as a person, and if they want to have sex with it they can?'

'Yes, that's correct.'

I have to ask him twice to be sure, because suddenly it all makes sense to me: the people who make sex robots are making slaves. Not human slaves, of course, but slaves who will one day be almost indistinguishable from humans. If they succeed, it will become normal for us to share our homes with beings we never have to empathize with, who exist only to fulfil our every wish, and who do everything human that most humans would rather not do.

Just as Matt, Roberto, Sergi and Steven have been trying to tell me all along, this really isn't about sex at all.

*

The sex robots of our collective imaginations – perfect synthetic companions without human flaws – don't exist. But they will do, and sooner than most of us realize. Within a decade or two, the technology will be advanced enough, and affordable enough, to make relationships with robots normal rather than niche.

The people who make these robots, and the academics and columnists debating them, are from a generation who will probably never have a relationship with one. Steven says most of the people who have paid the £300 deposit for DS's robotic head in Europe and North America are 'young men'. Paul Lumb, the British boss of Cloud Climax, the retailer with the licence to distribute DS Dolls in Europe, says the customers showing an interest in the dolls and robots are part of a new sexual revolution where almost anything goes. 'We've changed so much over the past ten years. We're so open regarding sexuality and sexual preferences,' he tells me.

Paul has warehouses in the Netherlands and the north west of England, and works with manufacturers all over Asia. He is constantly travelling. When I manage to reach him on the phone on a Sunday afternoon, he apologizes for being hard to find. 'It's an exciting time for us at the moment. It's business on steroids.' He talks like a contestant on *The Apprentice*, full of buzzwords and car metaphors. And he likes to talk; I don't have to ask him much at all.

'Personal self-gratification comes in many formats,' he says. 'To me, the dolls are the Bugatti Veyron of adult toys. They are a big investment – not just the financial investment but also the emotional investment. Not everybody's got the space and storage for a 168-centimetre, 38-kg doll.'

But when he talks about how much the videos of DS's robotic heads have caused a sensation on Instagram, he says something I'm not expecting.

'We are not lovers of social media, believe you me,' he

confides. 'It's probably altering people's psychology. We don't know if it has damaging effects on interaction and procreation – if you can't get in a relationship because you can only speak with your phone, how the hell are you going to start a family? It's that serious, really.'

'Do you not think that maybe the robots have the potential for that, though?' I ask. 'That you can get so used to being with a robot that you won't want to go out and meet a human flesh and blood person?'

For the first time, he pauses. 'You know, that's a deep question. It's a tough question.' It turns out it's one he doesn't want to answer. 'I know from a lot of our owners that the majority of them are in relationships. I wouldn't say we had anyone who was lonely and feels detached from society.' Not yet, at least.

'I'm quite old school, Jenny,' he continues. 'I'm forty-six years young. I think back to the days where there was no such thing as a mobile phone. I was moving around the country going to raves, and we were reliant upon the interaction with others through word of mouth and flyers to see the next DJ and the next venue. We experienced the summer of love, things that you can grasp and harness, things that build your character. We'd go into a bar or a club and we'd put the world to rest and we were very confident in the way we were going. That's been taken away from a lot of people now. Because of the world of technology, social interaction is more limited.'

But while I hear such loss in the changes Paul is describing, he sees commercial opportunity. 'Younger people work harder now, and longer hours, so downtime is at a premium. We're finding that there's an interaction with more technology-based products for distance relationships. You can have a relationship with someone you've built up through social networking and use all the products that are available for intimate distance relationships. That's something we really wanted to harness. We wanted to be at the forefront of technology. It's the next generation of lifestyle and well-being.'

Paul is right: since the turn of the millennium, there has been a proliferation of expressions of different kinds of sexuality and gender identity, a kaleidoscope of possibilities beyond the heterosexual that have been accepted and embraced like never before. It's a good thing, and we probably have technology to thank for it: social media has drawn people together, given them strength in numbers and a platform to speak to the world, and to each other, that could never have existed before.

But the same digital revolution has left us less prepared for face-to-face interactions, less able to relate in the real world, sexually liberated but socially stunted. It's normal to be Facebook friends with someone and follow them on Twitter but bury your face in your phone to ignore them if you happen to see them in your train carriage. Technology has isolated us, but our solution to our loneliness appears to be more technology. It's superficially alluring, but it makes no sense. Just like the earplugs in my Las Vegas hotel room, we are solving a problem with an extra layer of complexity instead of dealing with the cause of the problem itself.

So many of the arguments against sex robots focus on their impact on women, but the rise of the sex robot is going to affect us all. It's not just about the objectification of women – although the robots do objectify women. It's not only about men being given an opportunity to act out rape fantasies and misogynistic violence – although a small number may well want a sex robot for that reason. It's about how humanity will change when we can have relationships with robots. This is a humanist problem as much as a feminist one.

When it becomes possible to own a partner who exists purely

to please his or her owner, a constantly available partner without in-laws or menstrual cycles or bathroom habits or emotional baggage or independent ambitions, when it's possible to have an ideal sexual relationship without ever having to compromise, where the pleasure of only one half of the partnership matters, surely our capacity to have mutual relationships with other people will be diminished. When empathy is no longer a requirement of a social interaction, it will become a skill we have to work at – and we will all be a little less human.

PART TWO

THE FUTURE OF FOOD

Clean meat, clean conscience

CHAPTER FIVE

Cowschwitz

I smell it ten minutes before I see it. I've been driving up the Interstate 5 for three hours, and the desolate parched grass and brittle earth beside the highway is a repetitive and monotonous landscape, but the sharp stench of ammonia and sulphur – of piss and shit – jolts me to attention like a punch in the nose. By the time it swings into view I can feel it in my eyes, even though my car windows are shut.

A hundred thousand cows are crammed on the dull dust of the Harris Ranch feedlot – dust made up of the trampled manure of generations of cattle, baked in the California sun. Under a yellow haze stretching out to the horizon there are black cows, tawny cows, mottled white cows, clustered flank against flank, tagged ear against tagged ear, their tongues lolling, their legs caked in filth. They are not here to roam around; their sole purpose is to slurp grain, to quickly grow fat enough to turn into some of the 200 million pounds of beef produced by the Harris Ranch every year. As the cows throng too close together along endless steel troughs, they're no longer living creatures; they are items on an industrial production line.

This is the largest cattle ranch on the West Coast, and a hellish sight from my car window, but there are thirteen larger ones in the US alone, and it's small potatoes compared to the vast feedlots of Texas, Nebraska and Kansas, or the supersized dairy farms of China and Saudi Arabia. This window into the world of industrial agriculture is remarkable only because it's so transparent: located bang up against the highway, halfway between Los Angeles and San Francisco, there is nothing for it to hide behind. The Harris Ranch is notorious among American journalists, environmental campaigners and animal rights activists (some of whom destroyed fourteen tractors in an arson attack here in 2012). They prefer to call it by its nickname: 'Cowschwitz'.

I turn off the I-5 and head towards the Harris Ranch Restaurant and Inn, the affiliated high-end rest stop and shrine to beef. I check into a room filled with fat sofas upholstered in tan leather. The leather-bound room guide tells me I can order raw beef to take away, delivered direct to my door from the hotel's meat department. There is an inner courtyard with a turquoise swimming pool and Jacuzzi surrounded by empty sun loungers. No one is sitting outside and no one is standing on the balconies while the cloying, sweet smell of cow shit hangs thick in the air. Beef is on the menu for every course of every meal in the three restaurants here. You can start the day with a coffee-crusted ribeye, corned beef hash, the Breakfast Ranch Burger or smoked beef bacon. There are non-meat options, but diners are encouraged to 'Beef Up Your Salad by Adding Your Favourite Steak'.

I am no vegan. I like beef as much as the next carnivore – more, probably. For me, meat makes a meal, and steak is the king of foods, something to order on my birthday, the dinner my husband cooked for me the night we got engaged. I love the way it tastes, the way it feels in my mouth and in my stomach. And I eat it even though I know the meat industry is revolting, cruel, untenable, indefensible. Like the vast majority of the 95 per cent of the world's population who eat meat, I am happy to turn away from how meat is made, to shut my eyes when I open my mouth.

Veganism and vegetarianism might be more popular and accepted than at any other time in history, but those of us who do eat meat are consuming more than ever before. Take chicken: the amount of poultry eaten per person in the world's wealthiest countries increased by 50 per cent between 1997 and 2017. As the most populous countries get wealthier, they are becoming more carnivorous: in China, nearly twice as much beef was eaten per capita in 2017 than twenty years previously, and poultry consumption in India more than trebled between 1997 and 2017. The US alone eats 26 billion pounds of beef a year, which, converted into a stack of hamburgers, would stretch to the moon and back twice over, with plenty to spare. Yes, meat and dairy are good sources of protein, calcium and iron, but we're living in an age when we have the means and the knowledge to get the nutrition we need from plants and B12 supplements. Every year, 70 billion animals are killed for us to eat, not because they are good for us, but because we think they are tasty.

There are few worse things you can do for the health of humans, animals and the planet, for our earth, water, air and atmosphere, for the environment both within and outside our bodies, than eat meat. The evidence for this is unequivocal and monumental, and I'm sorry, fellow carnivores, but I'm going to break down exactly why.

First, climate change. The global livestock industry produces more greenhouse gases than the exhaust from every form of transport on the planet combined. The world's three biggest meat companies emitted more greenhouse gases in 2016 than the entire nation of France. The emissions come from animal feed production, the conversion of forests and grassland to pasture and cropland, and methane from cattle digestion (yes, cow farts). And we're talking about the very worst kind of emissions: methane is a far more dangerous contributor to climate change than carbon dioxide. For every 100 grams of beef, 105 kilograms of greenhouse gases are produced, not including those emitted when animals are transported to slaughter, or when their feed is transported to them, or the carbon dioxide they breathe out. If you add all of that together, as some environmentalists have done, then it's possible to argue that industrial agriculture is responsible for more than 50 per cent of global greenhouse emissions.

Second, medicine-resistant superbugs. In the UK, the NHS is trying to get us to take fewer antibiotics, because the more bacteria are exposed to them, the more opportunity they get to mutate into superbugs adapted to resist them. Got tonsillitis that feels like a medieval plague? Stick it out with some paracetamol, we're told. But that's a fat lot of use when 52 per cent of all the antibiotics used in China and 70 per cent of all those used in the USA are currently given to animals that aren't even sick. Antibiotics are routinely administered to make animals put on weight more rapidly and to prevent disease; animals crammed together with their own excrement, on top of the excrement of generations of others who have lived out their short, accelerated lives in the same tiny space, would be getting sick and dying faster than we could eat them if they weren't prophylactically dosed up. We might do things differently in Europe, but China and the USA combined produce twice the quantity of meat Europe does every year.

Without effective antimicrobial protection against infection, routine procedures like hip replacements, diabetes management, chemotherapy, organ transplantation or caesarean sections will become incredibly dangerous. Pneumonia and tuberculosis are already becoming difficult to treat, and the last resort of medicine for gonorrhoea (third-generation cephalosporin antibiotics) no longer works in at least ten countries, including the UK, France, Australia, Austria, Japan and Canada. If nothing changes, antibiotic resistance is predicted to kill ten million people a year by 2050.

Third, a carnivorous diet is a ridiculously inefficient way to put calories into your body. Instead of getting our energy from plants, we are getting it from animals that get it from plants. As well as producing the flesh that we eat, animals also make bone, blood, feathers and fur, they walk around and mate and chew or peck, they flap their wings. A lot of the energy they consume will never make its way to us. It takes thirty-four calories to produce just one calorie of beef, and eleven calories to produce one calorie of pork. The most efficient meat is chicken, and it still takes eight calories to get one calorie back out.

Fourth, water. The signs above the sinks at the Harris Ranch Inn and Restaurant might say, *During our **extreme drought conditions** please join us and limit your use of water,* but the management knows there are few more wasteful things you can do than use water to raise livestock. It takes 43,000 litres to produce the feed, drinking water and service water that ends up as one kilogram of beef – enough for a forty-eight-hour shower. If you think of it in terms of protein produced, you see how absurdly inefficient meat of all kinds is: it takes 112 litres of water to produce a gram of protein from beef, 57 litres to produce a gram of protein from pork and 34 litres to produce a gram of protein from chicken, but only 19 litres to produce a gram of protein from pulses. Hundreds of people have been killed in recent wildfires caused by drought that's become a normal part of life in California, but water continues to pour into the Harris Ranch feedlot.

And then there's water pollution. Outbreaks of *E. coli* and norovirus linked to salad and other vegetables are almost always traced back to the shit of farm animals contaminating irrigation water. Eutrophication, when manure and fertilizer leach into nearby water supplies and cause algae to grow and suffocate other aquatic life, has been found on 65 per cent of Europe's Atlantic coast and 78 per cent of the coastline of the continental US. We are killing fish when we eat meat.

Fifth, the sheer amount of land it takes to produce all that meat and dairy. Almost 80 per cent of all the planet's agricultural land is being used to graze animals or grow their feed, rather than to grow plants for our consumption. Up to 80 per cent of deforestation is estimated to be the result of agricultural expansion. Instead of being a vital asset for absorbing the carbon produced by animal agriculture, vast swathes of the Amazon have been burnt to the ground to make way for more grazing cattle and soya for animal feed. Researchers at Oxford University have calculated that if we were to stop consuming meat and dairy we could reduce global farmland by more than 75 per cent – equivalent to the landmass of the US, China, the European Union and Australia combined – and still feed the planet. We could use that land to grow trees or create solar farms or build homes or play laser tag: anything would be better than using it for industrial agriculture.

Sixth, meat gives us cancer, strokes, heart disease, obesity, diabetes, vCJD (the human form of mad cow disease), salmonella, listeria and *E. coli*. Eating animals is killing us.

So there you have six pretty cast-iron reasons why meat is indefensible, without a single mention of animal welfare, of how the vast majority of farm animals live short, horrible lives, and even those lucky enough to be treated well still have to die in order to satisfy our taste for meat. But you knew about all of that anyway. We might be able to ignore what our meat is because it's available in nice, sanitized, de-animalized packaging, but the undeniable fact is that eating meat is unjustifiable.

But it's also a fundamental part of human culture. To stop eating it would be to change the definition of a human diet, and to lose mankind's self-appointed role as master of the animals. One of the foundational pillars of human experience has become a threat to our very existence. There will be 9.7 billion humans living on the planet by 2050, and the Food and Agriculture Organization of the United Nations estimates there will be a 70 per cent increase in the demand for meat by that time. As much as most of the world's population would like it to, this can't go on without making the only known inhabitable place in the universe uninhabitable.

But, as well as being the birthplace of the Harris Ranch coffee-crusted breakfast ribeye, California is also home to the meat problem's most groundbreaking solution. Another three hours north along the I-5, a new wave of Silicon Valley entrepreneurs say we can carry on eating meat without any consequences, because they can produce it without raising animals. Not Quorn, or mock meat, or any kind of clever reconfigured plant protein that acts as a meat substitute; not the pea-and-coconut-oil-based Beyond Burger or the Impossible Burger that 'bleeds' fake blood. This is real meat, grown outside animal bodies: born in a flask, grown in a tank and harvested in a laboratory. The Silicon Valley start-ups are promising us flesh without the blood, meat unconnected to the land, meat that doesn't stink of shit, meat with a clean conscience. They are calling it 'clean meat'. And I've been invited to California so that I can be one of the first people in the world to taste it.

*

Lab-grown meat isn't a new idea (although not quite as ancient as Pygmalion). In his essay, 'Fifty Years Hence', first published in the *Strand* magazine in 1931, Winston Churchill ruminated on the direction in which scientific progress was taking mankind, and concluded that, by 1981, 'We shall escape the absurdity of growing a whole chicken in order to eat the breast or wing, by growing these parts separately under a suitable medium.' (This piece of writing has become so totemic in Silicon Valley that one of the venture capital funds that invests in food technology has named itself 'Fifty Years'.)

And disembodied flesh has been kept alive in laboratories since long before Churchill started thinking about it. On 17 January 1912, the Nobel Prize-winning French biologist Alexis Carrel prised a live chicken embryo from its egg and sliced a fragment out of its beating heart, then managed to keep the heart muscle tissue pumping away for over twenty years by bathing it in a special nutrient bath. When NASA wanted to find a way of producing fresh meat for consumption during extended space exploration, it funded the bioengineer Morris Benjaminson to conduct an experiment involving strips of goldfish meat, which were successfully grown in his laboratory in 2001. Benjaminson and his researchers cooked what they grew, but stopped short of eating it (although they did give it a sniff, and apparently it smelled tasty). The big shot in the arm for lab-grown meat came in 2004, when the Dutch government awarded €2 million to a group of universities in the Netherlands to research meat grown in vitro. But the funding ran out five years later, and it began to look like a pipe dream.

The world's first artificially grown hamburger was tasted at one p.m. on 5 August 2013, at a high-profile press conference for an invited audience of 200 journalists and academics in London. Made by the Dutch professor Mark Post, a physiologist at Maastricht University, the burger cost €250,000 to produce (around £215,000, or $325,000) and was bankrolled by Sergey Brin, Google co-founder and one of the world's richest men. The burger was a proof of concept rather than the beginning of a business, billed as 'the first recognizable meat product created using culturing techniques'.

It made headlines around the world that day. I saw it on the news, and the footage has stuck in my mind ever since. Professor Post unveils the burger by theatrically removing a silver cloche to reveal a puck of thin, pink squiggles of flesh in a Petri dish, 20,000 muscle strands grown in his laboratory (plus a little egg powder and breadcrumbs, he explains, and some red beet juice and saffron to get the colour right). A chef dressed in immaculate double-breasted whites fries it in a little butter, basting it regularly, and it's finally tasted by the food writer Josh Schonwald and food trends researcher Hanni Rützler, who pronounce it 'bland' and 'dry', but with a bite that had 'a kind of density that was familiar'. It wasn't quite right, but it was a triumph.

The launch was as corporate as it gets for an academic project, and an equally slick promotional film accompanied it. 'Sometimes a new technology comes along and it has the capability to transform how we view our world,' Brin says over reverberating guitar notes, managing to look both futuristic and completely dated in his Google Glass headset. 'I like to look at technology opportunities where the technology seems like it's on the cusp of viability, and if it succeeds there, it can be really transformative for the world.'

The video then cuts to Richard Wrangham, Professor of Biological Anthropology at Harvard University. 'We are a species *designed* to love meat,' he says. 'It's been *fantastically* beneficial for us. Once we started cooking meat, that enabled us to have lots of energy. That energy enabled us to have big brains and become physically, anatomically human.' It's OK to love eating meat, it's human nature, and it's *made* us human too. 'Hunters and gatherers all over the world are very sad if, for a few days at a time, the hunters come back empty-handed. The camp becomes quiet. The dancing stops. And then somebody catches some meat!' he exclaims, his fists clenched with delight. 'They bring the meat into the camp – or nowadays someone's back garden with a barbecue. Everybody gets excited.'

In the final half of the video, Post explains how the beef is actually grown. He makes it sound like a doddle. 'We take a few cells from a cow, muscle-specific stem cells that can only become muscle,' he explains. 'There is very little that we have to do to make these cells do the right thing. A few cells that we take from this cow can turn into ten tonnes of meat.' Piece of cake.

The reality is a little more complicated. A biopsy of stem cells is taken from an adult animal – they are called 'starter cells' because they have the ability to grow, divide and become fat and muscle (if you were to cut yourself, these cells would be the ones that allow the wound to regenerate). Only a very small number of starter cells are needed for the process to begin; a biopsy the size of a sesame seed is fine, and it can be taken from a live animal, under anaesthetic, if you so choose. The starter cells are put into a seed tray, bathed in a medium of nutrients and growth factors, and placed in a bioreactor to encourage them to proliferate. One cell becomes two, two four, four eight and so on, until there are trillions of cells. These are then organized into a gel scaffold that helps them form the shape of muscle fibres, which are eventually layered. It takes about ten weeks to grow enough cells to make a burger, but, because the growth is exponential, it only takes twelve weeks to produce enough for 100,000 burgers. (According to Mark Post, you'd get about 2,000 burgers from a single cow, who'd have to have lived at least eighteen months before slaughter.) Burgers, croquettes and sausage meat don't have much structure and are relatively straightforward to produce; a sirloin steak would require a serious amount of work to get the fat, cartilage and muscle into the correct texture and configuration. Just as advances in AI will accelerate because of the market created by sex robots, tissue culture technology will advance because of the potential to grow cuts of meat.

Unlike animal meat, clean meat can be fully dominated and controlled down to the last cell. The possibilities are potentially endless: meat with extra omega-3 fatty acids to counteract the heart disease caused by animal fats; meat without the risk of E. coli or salmonella, as no animal intestines are grown and no animal will shit itself with fear when it's being slaughtered (which happens on even the most friendly of farms); new textures and flavours and shapes of meat that would be impossible to create inside an animal; foie gras without force feeding; pigless, kosher bacon.

But none of this has been put on the market yet, even though there has been an explosion of start-ups around the world racing to be first. They've given themselves bucolic, wholesome names, like Mission Barns, Modern Meadow, Memphis Meats and Fork and Goode. It's the California entrepreneurs who are making the broadest strides, fuelled by the kind of investment that can only come from Silicon Valley venture capital. The meat and poultry industry in the US alone is worth over $1 trillion. Whoever can make a dent in that – even if it's only cornering 1 per cent of the market – is set to make billions.

*

I know all this because two weeks before I came to California I had a coffee on a drizzly day in London with a man called Bruce, who is neither scientist nor entrepreneur but is more responsible than anyone else on earth for the new clean meat industry. For two hours, Bruce leaned forward, his forearms on the table, and told me earnestly, intensely, unblinkingly, with a torrent of numbers, names and facts that he very much wanted me to write down, about how he has seen and tasted the salvation of the planet, and is on a mission to bring it to as many people as possible.

Bruce Friedrich is the executive director of the Good Food Institute, an American 'think tank accelerator' for the clean and plant-based meat market sectors. We met in a Mayfair cafe with loud monochrome floor tiles and overpriced flat whites because Bruce had just had a meeting round the corner with a British private-equity billionaire who is one of the GFI's most significant donors. Energetic and trim in a mint-green shirt, Bruce has piercing blue eyes that demand to be met. It was a week after the UN's Intergovernmental Panel on Climate Change had issued its latest warning that animal agriculture was the greatest contributor to greenhouse gas emissions, and there were stories all over the British press about how we have to stop eating so much meat. I was expecting Bruce to be happy about the headlines. He wasn't.

'Check back in eighteen months,' he said. 'Back in 2015, Chatham House argued that unless meat consumption goes down countries will not be able to keep climate change under two degrees by 2050. They got headlines then, but nobody paid attention. When the head of the IPCC, R. K. Pachauri, got the Nobel Peace Prize alongside Al Gore in 2007, he went, "Meat meat meat meat meat meat," and the UK media covered it extensively, and then, now, however many years later, people are like, "Oh my God, we've never heard of this."'

'Why is that?' I asked. 'Because people don't want to hear it?'

'The implications for people are beans and rice. People don't want to eat beans and rice. The fact that it happened last week doesn't mean that we're not going to be having exactly this same conversation in two or three years.'

'So there are cycles of selective amnesia?'

Bruce smiled. 'People are busy,' he said, generously. 'The entire thesis of GFI is, for decades we've been educating people about the harms of industrial agriculture, but education has not worked; 98 to 99 per cent of people are not going to meaningfully change their diets on the basis of environmental harm or global health harm or animal protection harm. The definition of insanity is doing the same thing over and over and expecting a different result. So give people what they want but produce it in a different way. Let's change food. Let's create meat directly from cells, without the inefficiencies and the antibiotic need and the cruelty of industrial meat consumption. Give people what they want, but remove the harm.'

It all sounded very free market and American. I thought of another Nobel Prize winner: Richard Thaler, who won the prize in 2017 for his theory of behavioural economics, of how to influence human behaviour by 'nudging' people towards the 'correct' choices.

But Bruce batted that idea away. 'It's even a little more elementary than nudge theory. This is the theory of the car replacing the horse and cart. If what people like about meat is the taste, the texture, the aroma – fairly fundamental things – if we can give them those things but in a better way, they will shift. If it's a better product and less expensive, people will shift.'

When the GFI was formed in 2015, there was just Bruce and one other staff member. Three years on, he was head of an organization that employed seventy people, in India, Brazil, Israel, China and Europe, as well as in the US. There was a single clean meat start-up, Memphis Meats, when the GFI launched;

there were at least twenty-five, three years later. This is largely down to how easy Bruce and his team are making it for entrepreneurs to start companies. The GFI has a science and technology department to publish peer-reviewed papers on clean meat research and development, an innovation department to help start-ups, a corporate engagement department to get huge food companies on board, and a policy department that lobbies governments to 'roll out the regulatory red carpet for clean meat', so that it can be labelled and sold alongside and eventually instead of meat grown in animals. Like the first ever lab-grown hamburger, the GFI is funded by tech entrepreneurs. Its biggest donor is Facebook co-founder Dustin Moskovitz and his wife.

Bruce goes into business schools and graduate science programmes to spread the word about clean meat to the next generation of entrepreneurs and researchers. The GFI publishes a ninety-eight-page manual, 'an all-you-can-read buffet on planning, launching, and growing a good food business', as it says on its cover – a kind of step-by-step idiot's guide to growing and selling meat from cells, with everything from how to hire a lawyer and get funding to search engine optimization and logo and packaging design that practically anyone could follow, free to download.

'Your start-up manual is quite something', I told him. 'It's very comprehensive.'

'Oh, thanks. We want everybody doing this. We would love to see environmental groups taking this up as a key aspect of their mandate.'

'But the environment doesn't really come into the manual. It's very slick, very Silicon Valley. It looks more like you're saying this is a fantastic business opportunity.'

'Oh, yeah. People are investing because they want to make a

lot of money and they see a global trillion-dollar meat industry and the possibility to produce meat less expensively.' He gives the same message at universities. 'We want everybody who's going to be the next titan of industry to be aware of clean meat as an outlet for their considerable talents. We want to find people who are tissue engineers, biochemical engineers or whatever and say, "Hey, you can be a part of saving the world and you can raise a family *very well* going into this space. You can simultaneously make money and self-actualize by doing something that can save the world from global catastrophe."'

There's that joke: How do you know if someone's vegan? Because they'll tell you. But it's not true in the world of clean meat. Throughout our conversation, Bruce only mentioned the 'v' word when I brought it up. He kept his former roles as director of vegan campaigns and then vice president of PETA quiet – he was open when I asked about it, but I doubt he would have mentioned it had I not. The Silicon Valley clean meat start-ups are run by vegans, and the people funding them are largely vegans too. The GFI itself depends on vegan money for its existence: Dustin Moskovitz and his wife happen to be vegan, and so is the British billionaire that Bruce was in Mayfair to meet. But Bruce didn't volunteer that either. For someone so forthcoming with facts and information, it was one of the few things I had to ask about. Clean meat was beginning to look like a vegan movement in disguise, one that knows the 'v' word is loaded with an attitude of moral rectitude that is toxic for those who love eating meat. But the future Bruce and the clean meat entrepreneurs are working towards is a world where the meat industry is owned and controlled by vegans. Clean meat is vegan meat.

When you're trying to wean human beings off animals without anyone noticing, language matters. After Mark Post first lifted the cloche on his burger in 2013, no one quite knew what to call his creation. Cultured meat? Lab-grown meat? In vitro meat? It was the GFI who did some serious market research and came up with the industry standard nomenclature. 'We coined "clean meat". We found that it had 20 to 25 per cent greater consumer acceptance than "cultured meat". I think people heard "cultured", and it meant Petri dishes.' The GFI advised start-ups to change their names for fear of alienating customers – like the Israeli start-up, now called Aleph Farms, which used to be called Meat the Future. 'People don't want futurism with their food,' Bruce declared.

The GFI wants consumers to focus on the end product rather than the process it takes to produce it. The meat industry does the same: after all, beef isn't called cow, and pork isn't called pig. Bruce said 'clean' meat invites parallels with clean energy and quickly conveys the idea that this meat is, by definition, free from antibiotics and pathogens. But if we all agree to call it clean meat, meat grown inside the bodies of animals becomes unclean, dirty. If we use the term Bruce wants us to use, we are quietly accepting the political position of veganism.

'People are going to know that this meat hasn't grown on its own,' I said. 'Surely that's going to turn a lot of people off?'

'I don't think we're going to have aaaaaany trouble with consumer acceptance. People eat meat right now *despite* not *because of* how it's produced. Show people an abattoir and ask them, "Do you want to eat this?" No. I think once it's produced in factories and it's streamed live on the internet, everybody will be down with it.'

'Production is going to be live-streamed on the internet?'

'Oh yeah. Absolutely. Transparency is critically important. A fully transparent process will put regulators' minds at ease, and the media's job is to be pessimistic and to play devil's advocate, so the coverage of it will be good if the companies are transparent, and dubious if the companies are not. But also, the people who are doing this are doing this for the right reasons. Transparency is baked into the cake.'

They are also doing it for the money, of course. 'If you capture only a tiny part of the global meat market, that's potentially a lot of money,' I said.

'But we're going to get *all* of that market,' he replied immediately.

Money is actually the last thing Bruce cares about. This became clear ninety minutes into our expensive coffees, when I asked him why he originally became vegan. It was 1987, he explained, he was a student, volunteering at a soup kitchen and organizing fasts in aid of Oxfam International (rather than getting drunk and eating kebabs, like most students I knew at university). Then he read *Diet for a Small Planet* by Frances Moore Lappé, a groundbreaking book from 1971, which argues that world hunger is caused by the inefficiencies of meat production.

'I thought, holy shit. I am basically living my life to try to eliminate global poverty and I'm eating meat, dairy and eggs; I'm eating foods that require way more calories than foods I could be eating. They're not particularly healthy, and they are leading to global starvation.'

'So you became a vegan because of human rights?'

'That was the reason I first went vegan. Then I went to work in a homeless shelter in inner city Washington DC for six years and I read *Christianity and the Rights of Animals* by Andrew Linzey. He's an Anglican priest.' Bruce fixed me again with his unwavering blue eyes. '*All* of this is based in my faith. My faith is *all* of this. The advocacy against global poverty was a recognition from Matthew 25 that salvation means you cast your lot with the poor and try to alleviate their suffering. Linzey's argument is that what's happening to other animals on factory farms *mocks God*. God designed animals to breathe fresh air and produce their young and give glory to God, and the way they are being treated on farms denies them everything God created them to be and to do, and inflicts pain on them for something so inconsequential as a palate preference. We're told that the earth is on loan to us – it tramples over that; that our bodies are on loan to us – we're dying of diseases of overconsumption. From a faith standpoint, it's wrong in every conceivable way.'

When he finally stopped for breath, Bruce had a serene, definitive smile on his face. The two things he had kept quiet during our conversation – his faith and his veganism – are clearly the motor that drives him and the centre of his universe. Given licence to talk about them, he switched mode and was suddenly evangelical, on a mission – a religious mission, an animal rights mission, a human rights mission – to save the planet, like some kind of vegan Christian superhero.

'Is this a calling for you?' I eventually asked.

'It is an absolute calling,' he replied, resolutely. 'It is a religious calling.' And there was something about his unapologetic earnestness, his sincere and intense conviction, that made me feel very cynical and English and carnivorous and small.

I wondered whether clean meat was vegan enough that you could eat it and still be vegan. 'You've tasted it,' I said. 'Do you still consider yourself vegan?'

'Yes. I think the fact that somebody eats meat three times doesn't mean that they're not a vegan. I don't think you can routinely eat clean meat and be a vegan, because clean meat is meat, and what a vegan does is not eat animal products, so once clean meat is widely available, I would stop being vegan, because I would eat clean meat.'

'What was it like to eat it, having not eaten meat for thirty years? It must have been weird.'

'I've now had chicken and duck. The first thing I thought was, Holy shit, this is good.'

Really? I thought. From everything I've heard from my vegan and vegetarian friends, if you haven't eaten flesh for decades and then suddenly try it, intentionally or not, the taste and texture are revolting, and it leaves you with horrible digestive problems.

'So you liked it?' I asked again.

'Oh yeah! I don't have an objection to the taste or the smell or the texture of meat. I have an objection to the external costs. But yeah, I loved it.'

Surely that's the thing. If our appetite for meat is killing us all, isn't the problem that needs addressing our desire for it, rather than the means by which the meat is produced?

'Isn't clean meat going to perpetuate the taste for meat among people who might one day go over to a plant-based diet, if we find a way to make the other arguments convincing?' I asked.

As ever, Bruce already had the answers. 'Three things,' he shot back. 'Thing one: that's about the biggest "if" in the world. We've tried before, and failed.'

'But aren't more people becoming vegan than ever?'

'When I first started advocating veganism professionally in 1996 I thought we were on the cusp of global veganism. There was this buzz. We had Alicia Silverstone, we had Alec Baldwin, we had Pamela Anderson, who were just huge in 1996. The numbers really haven't shifted much in all that time since.'

That's not what I had read about the growth of veganism in Great Britain, where the number of vegans is supposed to have quadrupled between 2014 and 2019, but global figures are impossible to come by, and surely Bruce's mastery of facts and numbers on this issue was bound to beat mine.

He was in his stride now. 'Thing two is just a colossal "So what?" Who cares? If you can produce meat from plants and directly from cells, then what's the objection to perpetuating the taste for meat?'

'Wouldn't there be some kind of black market for real meat – meat really from an animal?'

'It will be a tiny fraction of what it is now, and the animals will have lives that are worth living. If 100 per cent of the animals who were bred for food led good lives before they were slaughtered – which is what would happen in this eventual scenario – it would be well less than 1 per cent of the number of animals slaughtered now, and they would all be treated well.'

Before I could ask how he could possibly know this, he was already on his next point, the most important and the one that feels like the true reason for all his efforts. 'And then, thing three is – that probably goes away. In a world in which 98, 99 per cent of meat is plant-based meat and clean meat, that is a world in which the vast majority of people are not participating in the exploitation of animals on a daily basis. A big part of why animal rights has not caught on is that 98, 99 per cent of people are participating in cruelty that would be jailable cruelty, *every single day* –' he jabbed his index finger at the table with each word – 'if these animals had legal protection. If people are no longer participating in the cruelty on a daily basis, that makes moving towards a world in which animals are treated well, and their rights and interests are protected, a heck of a lot easier.'

So this is how the animal rights revolution is finally going to be won: not through horrific undercover animal testing videos, not through firebombing department stores that sell fur coats, but by giving us carnivores something in place of our meat that makes us re-evaluate the right we think we have to live at the expense of animals. To accept Bruce's position is to believe that the animal rights movement has failed, and that technology will lead to the changes that ethical vegan arguments have failed to convince us to make.

Bruce had a busy schedule: his next appointment was a meeting with KFC to discuss a chickenless future. I apologized for taking up so much of his time.

'There's really nothing I enjoy talking about more,' he said.

'I can tell,' I said.

When I'd set off to meet him that wet afternoon in London, I wasn't expecting to feel sold on clean meat by the end of it. But Bruce's confidence was contagious. There was no question I had to word carefully, no criticism I couldn't bring up, no problem to which clean meat wasn't already the solution. In his company, clean meat felt inevitable, a question of when, not if. At the end of it all, I felt as if I might have just had two hours with someone who is genuinely making history.

Two weeks later, as I check out of the Cowschwitz hotel and head north on the I-5 to San Francisco, I am still feeling Bruce's optimism. The Harris Ranch feedlot swings into my rear view mirror and disappears behind me, and ten minutes later the stench of it has faded away too.

CHAPTER SIX

The Vegans Who Love Meat

Tarpaulins flap in the breeze in San Francisco's Mission District. Tent camps nestle like moss, dark and unnoticed, against chain link fences. On Folsom Street, a homeless man is asleep, face down and sprawled across the pavement. A few paces away from his nose there is a golden door, polished so it shimmers in the midday sunlight. At its centre there is a glass panel bearing the word 'JUST'. This is the headquarters of the $1.1 billion food start-up that has announced it's about to become the first to sell clean meat to the public. It's 'JUST' as in 'guided by reason, justice and fairness', according to the strapline on its labels. There is nothing reasonable, just or fair about how venture capital billions can pour into a city alongside such manifest desperation, but the people who work around here don't seem to see it.

I am buzzed through the gold door, taken up some grey stairs and into a vast open plan office with a concrete floor. *Swoosh* – someone swerves around a bank of desks on a skateboard. Speakers play smooth jazz from somewhere up among the exposed steel beams and snaking pipes. Around a hundred people work here, but there are also two golden retrievers bounding around with wagging tails and lolloping tongues. A couple of kids are kneeling beside a low table, doing some colouring. Two enormous black and white photographs are framed and mounted side by side on one of the white walls. On the left, there's Bill Gates stuffing something into his mouth next to JUST CEO Josh Tetrick, with 'LEAP' superimposed in giant red letters in the bottom right-hand corner. On the right, there's Tony Blair also cramming something into his mouth as Josh looks on. This one is emblazoned with the word 'DARE'.

I'm here to taste the clean meat JUST is about to launch, and to meet Josh himself. But first I must have the JUST tour. 'The building was once a chocolate factory, and then for a while Disney Pixar were here,' communications manager Alex Dallago tells me, conjuring up the sense that dreams have long been confected in this space. Perhaps Josh is a kind of Willy Wonka who can make fantastical foods a reality. Alex won't tell me what I'm going to be eating today, though – it's going to be a surprise. I have to wait and see.

At least JUST is prepared to let me in the building. Despite all of Bruce's promises of total transparency and live-streamed production, it turns out the clean meat industry isn't very open to scrutiny at all – for the moment, at least. Memphis Meats – the first and biggest clean meat start-up – claims to have been producing beef since 2016, and chicken and duck since 2017. At the GFI's annual conference in 2018, Memphis Meats' CEO Uma Valeti said anyone who wanted to taste their meat was welcome to swing by their HQ and give it a try. But no journalist has tasted it yet, and all the images of their gently browned meatballs in a nest of fettuccine have been taken and distributed by Memphis Meats themselves. I'm sure they are culturing meat successfully – meat giants Tyson and Cargill, as well as Bill Gates and Richard Branson, have made significant investments in the company – but despite Uma's invitation they didn't want to share their creations with me. Their press officer kept giving me different reasons why it wasn't quite the right time: Uma was out of town, all the meat they had to taste was currently ear-marked for potential investors to try, they were renovating their premises and had no idea when the work would be finished – could be six months, could be longer.

JUST has had its own issues with transparency. When Josh founded the company in 2011 it was called Hampton Creek; its flagship product was a plant-based, eggless mayonnaise called JUST Mayo that became a commercial hit, outselling all the other mayonnaises, vegan and non-vegan, in Whole Foods. Hampton Creek's USP was that it scoured the world for plants that would yield proteins that could seamlessly replicate the properties of egg, using lab work and computational analysis to identify the perfect specimens. By claiming to be able to unlock the molecular secrets of plants – and to have hacked eggs, so they no longer have to come from chickens – Hampton Creek positioned itself as a tech company rather than just a vegan food producer, making it attractive to a raft of venture capitalists who would never invest in bean burgers. But in 2015, several former employees told journalists at Business Insider that 'the company used shoddy science, or ignored science completely' and that it 'stretched the truth' in order to attract those inves-tors. And in 2016, an investigation by Bloomberg suggested the booming sales figures of JUST Mayo needed to be taken with a pinch of salt; it found evidence that Hampton Creek employees and contractors had been instructed to buy up huge quantities of JUST Mayo from Whole Foods, massively inflat-ing the numbers.

In 2017 Hampton Creek was renamed JUST after its bestselling-or-not brand, and Josh decided they should branch out into clean meat, a completely different and even more high-tech area of both science and business. A new video went up on their website to explain the process. I watched it just before I arrived.

'We came up with the idea to use one feather from the single best chicken we could find,' says Josh's unmistakable, deep Southern voice, as the shot fades in on a lone chicken with fluffy white feathers, bathed in golden sunlight, on a broad pasture. A caption comes up over the bird, saying *Ian, Chicken*. A man in sandals stoops into the frame. He picks up one of Ian's discarded feathers from the grass and holds it up to the sunlight, twirling it in his fingers with an air of wonder as if he's just isolated the Higgs boson, and he deposits it in a transparent sample pot. Then there's a bit with some robots in a lab and some handwritten equations on one of those transparent boards you only ever see in sci-fi movies and forensic detective dramas. Science. At the end of the video there's a kind of outdoor cookout involving a deep-fat fryer, where a chef sprinkles sea salt with an exaggerated, slow-motion flourish over a tray of freshly fried Ian nuggets. Seven people sit around a picnic table, smiling with mouthfuls of Ian, while Ian himself struts around at their feet.

'It was an out-of-body experience to sit there and eat a chicken but have the chicken that you're eating running around in front of you,' Josh's voiceover says, even though he isn't one of the people eating Ian in the video. 'We've figured out how life really works, and now we don't need to cause death in order to create food.'

It's all unintentionally funny to my eyes, but Bruce's earnest sincerity has brought me here, so I resolve to park my cynicism and take this seriously. If even a part of what's being promised is correct, then our relationship with animals, the planet and our diets is about to change forever. And I could be among the first to experience it.

Before that, I have to learn about plants. Alex introduces me to Udi Lazimi, JUST's global plant sourcing lead, who will begin my tour. There is something in his scruffy beard and striking blue eyes that's strangely familiar, and it takes a few minutes of small talk before I register that Udi is the sandalled guy who held up the feather in the chicken video. But Udi's work has nothing to do with chickens.

'My job is to source plants from all over the world for our research,' he explains as he leads Alex and me downstairs and opens the door to the Plant Library, a huge, chilly room lined with floor-to-ceiling metal shelves containing huge plastic tubs. In the middle of the space, there is a table covered with a black cloth, on which someone has placed seven little white crucibles containing different seeds, for my benefit. 'There are over 2,000 varieties of plants in this collection,' Udi says proudly. 'The tour will take you through our Discovery Programme, starting here with the Plant Library, which is the first step, the upstream, where we bring in thousands of different materials and we feed it through the Discovery Pipeline to research the characteristics of these plant materials.' I am clearly not the first person Udi has taken upstream.

He tells me he's been to over sixty-five countries looking for 'protein-rich' seeds ('we're talking the Amazon, we're talking South-East Asia, East and West Africa, the Andean foot-hills . . .'), and while at first this conjures up images of Udi in a khaki pith helmet hacking away at the jungle with a machete it turns out he finds seeds by visiting markets. And while the samples on the table include Maya nuts gathered by indigenous people from the Guatemalan forest floor and seeds from a fruit grown only in the Colombian Amazon, there are also oats, ground flax seeds and powdered hemp seeds, which you can buy in the grocery store a block away from the JUST office.

'We require the seeds to be in powdered form to go through

the robots that you'll see next, downstream,' he tells me. Down-stream is back upstairs, where Chingyao Yang, JUST's associate director of automation, hands me a pair of goggles. 'Just for your safety,' he explains. 'When we step inside the Discovery Platform, the machines will be running experiments.'

The Discovery Platform is filled with banks and banks of devices. There's something called a Microlab Star, all blue lights and whirring pipettes. There are dispensing units enclosed in glass pyramids, with rows of little bottles bearing the JUST logo. There are two impressive white robotic arms in glass cases that remind me of the exhibits I saw in the Science Museum with Kathleen, except they aren't moving. 'We call them Randy and Heidi,' Chingyao smiles. 'They can show us the nuances of the protein isolates in terms of their functionality, in terms of gelation or emulsification.'

I can just about work out what's supposed to be going on behind the jargon: this is where they analyse the seeds' proteins, looking for properties like melting temperature and viscosity, which they send on to the product developers and chefs who make JUST mayo, cookie dough, salad dressing and so on. There are supposedly a dozen scientists and engineers on the Discovery team, but I can see only one person working, and she is doing something manually with a single pipette. Do they really need all this technology?

'How often do you use these machines?'

'Most of them are running twenty-four seven,' he replies.

'So is this running at the moment?' I ask, pointing to the nearest giant gadget, which looks complicated and expensive, but is still and silent, like almost everything else in this space.

'Right now, no. They just went to a meeting so there's no sample, but typically there's always samples here. Right now, it's on standby.'

Alex leads me back downstairs, where Vitor Santo, the senior scientist of what JUST are calling their 'Cellular Agriculture Platform', is waiting for us in a corridor. Vitor is a tissue engineer; he spent five years working in cancer research before he moved from Portugal to San Francisco a year ago to work for JUST. He extends his long, slim arm to shake my hand, and then immediately launches into his part of the tour. This is the bit of JUST I really want to learn about, but, like Udi and Chingyao, Vitor has prepared what he is going to say to me, and he wants to deliver his lines regardless of what my questions might be, or whatever I might already know.

'You start with a small isolation from cells from the animal, like a biopsy. You take this to the lab and you culture the cells with nutrients, a liquid medium that contains all the things the cells typically need,' he begins.

'How much can you tell me about all of that?' I ask. 'How do you get your biopsies? What's your medium?'

'I can tell you that we are working with different species. Our most advanced product is chicken, but we are also working with beef, pork, other avian species. In terms of the media, we are following in the footsteps of general pharmaceutical and medical research that typically uses a lot of these media *recipes*, let's say. But we're making modifications in the composition of the media in order to make it, erm, more affordable.'

Vitor is choosing his words carefully, because the medium in which the cells are grown is a very big deal. Pharmaceutical and medical researchers like to use foetal bovine serum (FBS), which, as the name implies, is made from unborn baby cows. Serum is blood without cells, platelets or clotting factors, but it has nutrients, hormones and growth factors that cause cells to proliferate. FBS is extracted by plunging a needle into the beating heart of a living calf foetus that has just been sliced from its mother's uterus in an abattoir. Blood is drained from its heart for about five minutes until the foetus dies, and the serum is then extracted. It is difficult to imagine a less vegan substance than FBS.

But serum is really good for growing cells. Serum from calf foetuses is particularly rich in growth factors, and FBS is a universal growth medium, meaning you can chuck practically any kind of cell in it and it will flourish and proliferate. Other media exist, but they tend to work for only one or two specific cell types, whereas you can use FBS to grow whatever you like. It's been an important part of medical studies, used to develop vaccines, and in cancer and HIV research, and it was the juice Mark Post used to grow his famous hamburger. It's also a major reason why his burger was so absurdly expensive: FBS costs anything from £300 to £700 a litre, and it takes fifty litres of the stuff to produce a single burger.

'If we used the conventionally used formulas we'd never be able to release an affordable product,' Vitor continues. 'Our strategy here at JUST is to use our Discovery Platform to test different plant-based proteins and check which ones promote cell growth. We are actually going to feed the animal cells we've isolated with plant-based protein. If you think about it, that's actually what happens in nature: the animals feed themselves with plants.' This is oversimplified, of course – the medium is more than just food – but if JUST have managed to do this it will be quite a selling point. As well as being the first to put clean meat on the market, they will be doing it in the most vegan-friendly way possible.

'Have you actually found a plant-based medium that works?' I ask.

'So. I would say there is still work in progress; we still are screening a lot of the plants,' he replies. 'We have some formulations that work pretty well, but I wouldn't call it our final recipe. What I would say is that we have an animal serum-free medium.'

Even if JUST manage to grow meat on an industrial scale without animal serum, they will still need animals for the starter cells. I wonder how realistic a business model based on Ian feathers can be.

'Which cells do you use?' I ask.

'You can get them from a piece of muscle, you can get it from blood – it really depends on the animal. I can't give you a lot of detail on that. That's part of the IP.'

'Can you get it from a feather?'

'Yes,' he says, with a slight shrug.

'Of course. Because it's in your video.'

Vitor hands me my second pair of goggles of the tour and we step into another lab. In the far corner there are three metal extractor hoods over seed trays, little plastic matrices with tiny wells of bright red fluid containing meat cells in each space, and a woman is pipetting something into them. Vitor tells me she's changing the medium; it needs to be constantly refreshed because the cells consume nutrients from it and leave waste materials in it that would inhibit growth if they weren't removed. He says this all in a breezy way, as if the process were just gardening in lab coats.

Everything in this room is here to trick the cells into thinking they are growing inside an animal so they reproduce. The regularly refreshed medium takes the place of the nutrients and waste removal they would get from blood pumped by the heart. The four grey incubators keep them at a body temperature of thirty-seven degrees. There's even an agitator, a moving platform, which gyrates the liquid suspension of cells to mimic what they would experience if they were growing inside a moving body. The swirling conical flasks of medium and meat juice look like something straight out of science fiction, but Vitor is keen to dispel that idea. 'It's very common, actually. It's used a lot for bacteria fermentation processes. If you think about beer, this is the same,' he says firmly.

They identify the most promising-looking cells in this lab, which then get taken upstairs to be produced on a larger scale in bioreactors and are finally sent to the JUST chefs for product development. 'From one chicken like Ian, we might have enough cells for the whole thing. We create this cell bank, thousands of little vials, and every time we start the production line we just take one vial and start from there.' Vitor smiles proudly. The idea that the thousands of crammed animals in the stench and filth of the Harris Ranch feedlot can be replaced with shelves of sterile vials is a remarkable one.

Alex, the communications manager, is a constant presence, nodding along to our conversation while checking her phone. She wants us to move on, to go back upstairs to see the bioprocess and production lab. I just want to taste the meat now. I wish they'd tell me what I'm going to eat.

'How far off are we from growing cuts of meat?' I ask.

'We could grow a steak in a week, if we wanted to,' Vitor replies airily.

This stops me dead. 'Really?'

'It's a matter of doing it in a scalable way. We could do a lot of prototyping and show the potential of the technology, but we don't. We know how to do it, it will just take a while to integrate into this workflow.'

If growing tissues is so easy, why do burn victims need painful skin grafts? Why are so many people on dialysis? Why aren't we just growing the kidneys and livers and corneas we need in labs, instead of waiting for people to die and donate them? Like so much of what I've heard today, it raises so many more questions than it answers.

The upstairs lab is bright and sparse. The two metal bioreactors are each the size and shape of a hotel minibar, and neither of them is running today. JUST have promised to release their clean meat on the market this year, but it's November; there is no way they could be mass-producing it from this room, from these machines. This looks like a research project, not the beginning of a commercial production line.

'When you are in full production, you'll need much bigger bioreactors, won't you?' I ask.

'That is correct. In order to reach the scales we need, we will need to build the bioreactors from scratch. It's a challenge. That's why it's important that we release the product and people can really taste and see its potential. Because once we get their support and finance from meat companies or other investors, then we can work on it.'

And then I realize that JUST has no intention of selling clean meat in shops any time soon. The launch will be a publicity stunt that means they can claim the title of being first and attract more venture capital. Clean meat is still at the proof of concept stage, although the concept being proven this time is not that meat can be grown in a lab, but that people are prepared to pay for it.

'How much is it going to cost?' I ask.

'Right now, I don't have an answer. It's going to be available in a few high-end restaurants; a limited release, this year.'

'Definitely this year?'

'Yeah. In one month or so, everyone will know about this.' He beams with confidence, and pride. 'It's incredible. It's been one of the reasons why I shifted from medical research to this project, because I felt that whatever I was going to be doing was going to have such a tremendous impact and would happen really fast. In medical research it takes, like, fifteen years to actually get a drug out to market. The way this industry operates is faster, and I got into it at the right time, when the right support was there.'

If you are eager, idealistic, ambitious and impatient, JUST is just the right place for you.

*

Alex leads me back to the open-plan office. 'Take a seat,' she says, gesturing towards a long black table where customer service manager Josh Hyman is waiting for me in front of a camping stove, wearing a grey cap and a black JUST apron, as if he were on the set of a home shopping channel or cookery show. Two hours into the tour, the time has finally come to taste the future. I'm absolutely psyched.

'Any allergies, sensitivities, things you do not eat?' he asks me as he fires up the stove. He should know this already; I had to email Alex my dietary requirements before I arrived. Of course, I don't have any. I'll eat pretty much anything, which is why I'm here. I'm trying not to be cynical, but it feels like they want to suss out my veganess in advance, to see how familiar I am with what I'm about to taste.

It turns out I'm still not getting to eat the meat yet. Not just yet. First, I have to try JUST Egg – which is eggless, of course – one of their plant-based creations from the Discovery Platform.

Josh scoops something out of a jar and it sizzles in the pan.

'Is that real butter?' I ask.

'Yeah,' he replies, casual as anything. 'I figure that 95 per cent of the people that eat scrambled eggs do this. So, why not be like them? It doesn't hurt. Makes it taste good.'

'What? This is a vegan company, a food company built on the promise that it doesn't exploit animals, and you're telling me butter doesn't hurt and makes things taste good?' I want to say. But I don't.

'It's fat,' he continues, breezily. 'I could use oil, but I don't want to. That's why I asked if you were allergic to anything. Are you ready? Here it goes. Mung bean egg.'

He pours out the JUST Egg from a 12 fl oz plastic bottle into the hot pan. It's pale yellow and shiny, just like a freshly whisked egg. It bubbles and sizzles, like an egg would. It begins to brown around the edges, puckering and curling a little, just like an egg would. It's kind of incredible that this *isn't* egg.

'You can even flip it, no problems.' He turns it over with a spatula. 'I'm going to use two things to season it with.' He takes a pinch of something from a grey crucible. 'The first is something called black salt. It's not 100 per cent necessary, but it has a naturally occurring sulphur compound to it, so it just gives you a *little* bit of that eggy smell and that eggy flavour. Just a *liiittle* bit. And, because it's eggs, I'm going to do a little bit of cracked pepper too. And that's really it. Looks done to me.' He serves it out into a bowl and hands it to me.

This looks like egg. It sounds like egg, it cooks like egg. It feels like egg on my fork and in my mouth – fluffy and spongy and hot. But it is totally bland. Without butter and pepper and special sulphurous salt, it would taste of nothing at all.

'It's pretty good,' I say.

'Right? It has a little bit of a snap, a little bit of sponginess, not too much.'

I don't know what else to say. 'It's good . . . It's *different*.'

'Yes. And, while it doesn't taste *exaaaactly* like an egg . . .'

'The texture is the same,' I say, trying to be constructive.

'The texture is really good. So if you think about it in a

system, if you think about it sautéed with vegetables, or if you add cheese to make an omelette, or put it in a breakfast burrito . . .'

In other words, it's fine so long as you completely mask what it is, or what it isn't. If this is the cutting edge, the apex of plant-based food technology, I can see why there's a need for clean meat. Despite all JUST's exotic seeds and clever robots, they still can't turn plants into animal proteins.

'Now for what you really came for,' Josh declares, producing a black dish out of nowhere. 'There's our little nugget.'

And I look down to see a small, lone, beige-crusted rectangle nesting in an envelope of greaseproof paper with red, white and blue stripes. An all-American chicken nugget.

'You can dip it in a little bit of the sauce, if you like,' he says, pointing to a small metal bowl of something pinky-yellow that sits on the plate beside it.

'So it's already cooked and done?' I thought he was going to fry something up on the stove for me. This feels weird.

'It's cooked and done,' he nods.

'And what's the sauce?'

'I believe it might be a little of our chipotle ranch.'

'I'm going to have it without the sauce first,' I say.

'You do your thing.'

'OK. Here we go.'

I bite through the batter. It's warm, crispy, deep-fried, heavily seasoned. Then there is the meat. And, yes, it is chicken. It tastes like a chicken nugget: there is the flavour, the aroma of chicken on my tongue and in my nose. But it's so mushy. So very, very mushy. Yet – chicken.

'Tastes like chicken?' Josh asks immediately.

'Tastes like a chicken nugget,' I reply.

'Yup!' says Alex, triumphantly, and they both beam.

As I continue to chew, I gradually realize that it is disgusting. At first, the meat is familiar – it has the juiciness, the unmistakable tackiness of animal flesh on my teeth – but it has the texture of the most low-grade processed food I could ever imagine. The consistency is so wrong, the meat is so far removed from animal tissue, that my brain is telling me this is very bad meat indeed and I should spit it out. There are no discernible pieces of meat in this nugget. It is chickeny mash, bulked with filler, inside a crunchy crust.

Josh fills the silence. 'Be critical, if you want to. We take any and all feedback.'

'The inside is a little bit . . . It's a bit mushy.'

He nods. 'OK.'

'What else is in here to make the nugget?'

'We combine, like, a few plant products, and then we add the cells to them. Other than the cells, it's a completely plant-based nugget.'

'How much of it is the actual meat?'

'Er . . . that I do not know.'

'So you didn't make this nugget?'

'I didn't, but Nicholas, who is just behind me, did.' He gestures to some people working with their heads down at some benches several metres away. I can't see which person he means. Nicholas is not on today's tour itinerary.

The nugget is small. There's only enough for three bites, and I have to nibble to make it last. I have no idea what I'm eating. This is an even more unsettling experience than meeting Harmony; at least I got to see how Harmony was made, but I've had a two-hour tour here, and I haven't seen any raw meat. The nugget arrived warm, but I didn't see it being cooked. I so wanted this to be a chicken nugget, but it doesn't seem to be at all.

Then again, I haven't had nuggets since I was a teenager.

What do I know? Maybe they are all this processed and mushy. Maybe this tastes exactly like it should. But perhaps Josh has no idea how a chicken nugget should taste either.

'Are you a vegan?' I ask.

'Er . . . yes,' Josh says, and he goes truly red with embarrassment, as if I've just discovered he's a secret nudist.

'Would you eat clean meat? As a vegan?' I ask.

'I've tried it, so obviously the answer so far has been yes.'

'Have you been vegan a long time?'

'Ten years. But I'm not super sensitive about it. This might be a way that I can jump back in the meat-eating game and not feel guilty. I'm not friends with a lot of vegan people. My wife isn't vegan. It was a choice I made myself, for me, not for anybody else. No offence to anybody, but I don't really care what anybody else does.' He's almost contrite when he says this, desperate for me to know that he's not part of a cult, that he's not judging me.

'Is it easy to cook with, if you're a chef?'

'Luckily, I'm not a chef. I'm front of house. That's why I'm here, talking to you.'

When I was in the RealBotix workshop, I saw a demonstration. At JUST, I'm getting a performance. I was given my first taste of clean meat by the front of house guy. The tour was choreographed and carefully stage-managed by Alex. So much has been simplified, romanticized and sidestepped to gloss over how far clean meat is from being ready for human consumption. I have no idea if what I just ate was grown in baby calf blood or magic plant juice. I don't even know which part of the chicken it originally came from – blood? Bone? Feathers? This has been an exercise in spin, an entertaining one, and one that I know will go far; the story of JUST's adventures in meat is a story journalists will want to tell, and investors will want to hear. But it's a story.

I thank Josh and Alex. 'This is a really big deal,' I say, 'because, if you get this right, the potential is so enormous.'

'I know. That's why we're doing this,' he smiles. 'We don't do little things around here. Josh Tetrick isn't into doing things that have minimal impact. It's either worldwide impact or nothing.'

I take a big gulp of water. I need to rinse my mouth.

*

One final act of the show remains, of course: Josh Tetrick himself, the founder, CEO and undisputed boss of JUST. Three Hampton Creek executives were fired last year amid rumours they were plotting to take control away from Josh and hand it to investors. A few weeks later, everyone on the board resigned, apart from Josh. Josh rules, and it seems like anyone who doubts it is out the door.

He's in his late thirties, as broad-shouldered as an American football player, with big hands and thick eyebrows. As I take a seat next to him at the meeting table I'm desperate for some genuine spontaneity, some unvarnished straightforwardness. If anyone can give me definitive answers around here, it should be this man. But Josh, too, has lines to deliver. When I ask him why he chose to go from making plant-based egg to clean meat, he has paragraphs at the ready.

'We're not a plant-based company or an animal-based company, we just want to be an effective company,' he says, in that deep Southern accent. 'It turns out that the mung bean is really effective in helping us make the egg, but if we really want to make beef, if we really want to make pork, if we really want to make chicken, we think it's more effective to start from a cell from a cow, from a chicken, from a pig, from a taste perspective, a texture perspective, and also a naming perspective.'

Josh knows all about the importance of naming things

properly. Unilever, which owns the Hellmann's brand, brought a lawsuit against Hampton Creek in 2014 claiming the JUST Mayo name was false advertising: instead of being 'just' mayonnaise, it was 'not' mayonnaise; it couldn't meet the definition of mayonnaise given by the US Food and Drug Administration, because it didn't contain eggs. The FDA agreed, and Hampton Creek changed their labels in 2015 to clarify what the product was, adding the 'guided by reason, justice, and fairness' strapline to show the specific way they were defining the word 'just'. Hampton Creek still got to call it mayonnaise, and people could buy it without thinking they were getting something alternative and strange.

'My parents shop for meat at Piggly Wiggly and Winn-Dixie in Birmingham, Alabama. How do I increase the probability that my mom's friends buy the kind of beef and the kind of pork I think they should buy, the kinds that didn't require killing an animal or using the land and the water? If it's not called meat, you're not going to create a system in the future where the majority of pounds of meat produced in a given day does not require an animal. That's the day I want.' His face is as fervent as a pastor preaching about the Promised Land. 'How do we accelerate the day when fifty plus one per cent of the meat produced was created without needing to kill a single animal? Because, when that day happens, the next day will be fifty-five, and then sixty. This is the *only* way to get to that day.'

For clean meat to work, it has to have mass appeal. There's no point being the bestseller at Whole Foods or Waitrose if the vast majority of people shop at Walmart and Tesco. This has to be about basic food – staples, not indulgences.

'Our end point is to change a system and along the way help our investors make a lot of money. Because I want them to invest more in us,' he tells me. He has eye-catching ideas about how to make this happen: as well as being the first on the market, JUST is going to turn the most expensive gourmet delicacies into basic food for everyone.

'We want to focus on the Kobes, the Wagyus, the bluefin tunas. I imagine my dad or mom walking into that Piggly Wiggly, and I imagine them looking at two kinds of hamburgers: one kind just says, *Ground chuck, $2.99 per pound* – that's what they've always been buying; another one says, *Kobe A5 burger, Wagyu A5 burger, $2.49 a pound.* One was made by killing the animal, the other was made by, you know, this different approach. I want my dad and my mom to say, "Well, this is obvious; of course I will choose a hamburger that is richer, more delicate, that hits you in the face with flavour, rather than the ground chuck." To me, that is what is required to create the different system.'

He flips open his laptop. 'Our plan is to have this out before the end of next year.' There's an image of two burgers on a white polystyrene tray with a red label saying, *2 Kobe A5 Beef Patties, 100% Japanese Wagyu.* The burgers are made up of chunky pieces of meat, heavily dappled with fat.

'A proper bit of marbled Kobe beef?' I ask.

'It will be Wagyu. Kobe is a form of Wagyu.'

I am taking lessons in beef from a vegan.

There are more concept pictures: two plump chicken breasts, some shimmering slices of deep pink bluefin tuna ('finest grade Otoro'), and then some drawings of the JUST clean meat factory of the future, complete with forty-eight separate 200,000-litre bioreactors, each the size of a power plant cooling tower, greenhouses for growing the plants they will use to make the medium, and a viewing platform where members of the public can watch the tuna steaks and chicken breasts being assembled on conveyor belts.

For this farm of the future to become reality, Josh says they will have to work with the meat industry, which already has the refrigeration and distribution networks JUST will need to bring clean meat to the masses. 'They don't want the chickens. Who wants to deal with 400,000 fucking chickens in a giant facility, shitting and pissing all over the place? If there's a better way to convert things into dollars, of course they'll do it.' Clean meat will become the only thing on the menu if it's cheaper and better for consumers, and easier and more profitable for producers. Market forces will save the planet. And then JUST will take over the meat industry.

'Are you aiming to be the biggest meat company in the world?'

He looks me dead in the eye, and nods slowly. 'Definitely.'

But first there is the matter of the launch. It's going to be a very small launch, before the end of the year – chicken nuggets in a couple of restaurants outside the US, he tells me. 'We're talking to a handful of different countries about the right place to launch it. The regulatory is not ready in the US,' he sighs. 'Politics, politics.'

That's one way of looking at it. Another is that he's trying to find a country with more flexible public health standards in which to experiment with his mushy nuggets.

'Will that be an ongoing thing or a one-off event?'

'It will be ongoing.'

'How much is it going to cost?'

'I don't know yet. It's still uncertain.'

'Do you know how much it cost to make the nugget that I just ate?'

He shakes his head.

'A lot of money?'

'Yup.'

'I just ate something very expensive?'

'For sure.'

'In the hundreds of dollars or the thousands of dollars?'

'Somewhere in the hundreds, but I'm not sure exactly. Part of our unknowing is that it doesn't even make sense for us to calculate the economics of it right now, because we still need to scale it up.'

Josh is suddenly unforthcoming. The paragraphs are gone, and I'm having to draw answers out of him sentence by sentence. So I change tack. Perhaps he'll feel more comfortable talking about himself.

He tells me he grew up in Alabama thinking he was going to play linebacker in the NFL, but realized he wasn't good enough when he got to college. He spent some time working with the UN Development Programme in Kenya, and got a fellowship to work with the investment ministry in Liberia, where he saw desperate poverty first-hand. 'I just got frustrated with governments and non-profits. For me, everything was just taking *too long*. So I got back to the US and I was like, "How do we increase the percentage of people who are eating well?"' He's back in preacher mode. 'Eating well, to me, means eating in a way that doesn't require killing an animal. Eating well means eating in a way that's more restorative to the environment. Eating well means not fucking up your body. Eating well means it's got to taste fucking good. Eating well means I can afford it. How can we increase the number of human beings who are eating well tomorrow? That's really the mission of the company.' A very broad mission indeed.

Josh went vegan ten years ago, but he doesn't go into detail about it. 'I would prefer to cause less harm in my meals. That's all,' he says, simply.

'Where does your sense of morality come from?' I ask, with

Bruce on my mind. 'Is it from an animal rights perspective? A human rights perspective? Is it a religious thing?'

'Yeah, no. To me, the more we can create a system where living things are flourishing, the better. That's my morality.'

'But where does that come from in your background? That's a very San Francisco way of looking at things.'

'I don't know. It's hard to tell, honestly.'

'I'm just trying to work out where you're coming from. When you were growing up, you didn't think you were going to one day grow meat in laboratories, obviously.'

'I do have to say, when this scales up, it won't be grown in laboratories. Yoghurt started up in a lab until Danone or whatever started producing a gajillion tons of it.'

This is total bullshit, of course. Humans have been making yoghurt for thousands of years. In caves. But I don't want to say so, because Josh is starting to get fed up with me, and I have one more question.

'Shouldn't we all just be eating less meat, rather than going to all these great lengths to make it?'

'Yeah, in the same way that we should all be walking to work instead of taking cars, in the same way that we should all be swimming across the Atlantic Ocean instead of using a jumbo jet. In the same way that we should all be growing our own crops instead of going to the grocery store. Yes, we should, but we have to live in a reality-based world.'

Josh doesn't live in a reality-based world. He lives in San Francisco, in the fake-it-till-you-make-it start-up culture where problems are obscured and outlandish claims are made with unshakable confidence so that all-important venture capital can be secured. When I look at the JUST concept images, I see a shiny idea to attract investment instead of a workable solution to the crises caused by the human appetite for meat. If this is what the rest of the clean meat industry is like, a few people may make some money in the short term, but all of us – our planet as well as our bodies – will pay the price of allowing business as usual to continue.

CHAPTER SEVEN

Fish Out of Water

On the day when the San Francisco Bay Area has the worst air quality in the entire world, I am in Emeryville, on the other side of the water from JUST. California wildfires that even the most stubborn climate change denialists accept are linked to a change in the climate have already claimed the lives of more than a hundred people, and the ash fog is so thick I can barely see across the street.

I haven't been able to eat meat since I ate the JUST nugget four days ago. The thought of it makes me nauseous. Perhaps clean meat is going to make me vegan after all, for all the wrong reasons.

My mind is as unsettled as my stomach. Have I come all the way over here just to see a Silicon Valley bubble, a stunt with no viable product to sell? Is the JUST chicken nugget the Roxxxy True Companion of clean meat? I'm still hungry for the authenticity and transparency Bruce had promised me.

So I'm pleasantly surprised when I ring the bell of Finless Foods and the company's CEO answers the door. Mike Selden has close-set eyes and a neatly trimmed beard. He's tall – six foot three – and humbly hunches over to shake my hand. I immediately know I'm in the company of an unassuming nerd.

He pulls his CSO and co-founder Brian Wyrwas out of the tiny boardroom so he can meet me too. They are East Coast natives who moved here from New York two years ago to grow fish. Brian is twenty-six and Mike is twenty-seven. 'We're the two youngest people in the company,' Mike declares. 'We share a company, we share a house, we share a car and pretty much the entirety of our friendship group. People assume that we're married. We don't do much to dispel that myth.'

Founded in 2017, not long after Mike and Brian had finished their biochemistry degrees, Finless Foods was the first clean meat start-up to specialize in seafood. They are focusing on bluefin tuna and sea bass; whatever they sell is going to be expensive at first, so it needs to fit the bill. Brian is friendly, but keen to get back to his meeting; they are deciding which of Finless's seven staff members are going to fly out to Asia to fetch some bluefin starter cells.

Mike is the 'front of house' in this start-up, but I am not going to see any kind of performance here. There will be no tasting. 'We did a bunch of prototyping and tastings earlier on, but a lot of that was just to . . . You need to play the investment game,' he says, with a knowing smile. 'Investors need to see something physical, which I think makes sense. Business is about feelings. There's plenty of brilliant scientists out there making companies who can't get any funding because they can't play the game.' But the foods made by Finless are not yet ready for market, and Mike isn't going to pretend otherwise; he is a scientist first and an entrepreneur second, with an academic's unwillingness to overplay his hand, in case he gets slapped in the face.

There are only three clean meat companies that focus exclusively on fish at the moment, which is surprising, given that the fish problem is more pressing than the meat problem. If meat is murder, fish is genocide. Decades of commercial fishing, using ever more voracious methods of catch, have led to an ecological catastrophe in our oceans. A third of all fish stocks are being depleted faster than they can ever replenish; that means they are so overfished that the population can't recover and the food chain has been destroyed. Another 60 per cent are already being exploited to their fullest extent – we cannot get any more fish out of them than we already do. That leaves only 7 per cent that are underfished, and these are often in areas that are too far from land to make it financially viable, or they're in politically contested areas (where you risk starting a war if you sail into them). In other words, we've pretty much taken all the fish we can get from the sea.

Fishing fleets are having to sail further out to catch fewer and smaller fish, burning more fuel. And yet 40 per cent of what commercial fishermen catch is thrown away – it is 'bycatch', the unintended, unwanted fish, turtles, birds and marine mammals caught in nets, killed and then discarded. We eat more fish than any other kind of animal protein and a billion people rely on it as their protein source. Poor coastal communities that depend on subsistence fishing are feeling the effects of this ecological catastrophe more than any of the rest of us.

Fish farming might sound like a solution to the destruction of ocean ecosystems, but it runs into the same problems as intensive animal agriculture. Large numbers of fish enclosed in a small area means a huge tank full of shit, and it requires pesticides, fungicides and insecticides to kill the sea lice that thrive in these conditions. And many fish just won't survive in a tank. Bluefin tuna need to move about a lot; being packed together like sardines in a tin will kill them.

So it feels a little naive to ask Mike why he chose to make fish rather than animal meat, but that's where I begin.

'There's a million reasons', he enthuses, delighted to be asked. First, our consumption of fish is 'the largest source of suffering on the planet. If you kill a cow, it can feed, like, 300 people, but if one person is eating fish, if you're eating sardines, you're eating, like, ten animals. It's suffering and killing on a much more massive scale.' Then there are the health reasons. 'With bluefin tuna, there's mercury and plastic. The EPA-FDA [the US's Environmental Protection Agency and the Food and Drug Administration] recommend that women of childbearing age – which they have decided is between the ages of sixteen and forty-nine, their numbers, not mine – shouldn't eat *any* large carnivorous fish *at all* because of the mercury. Other people should only eat it once a week. Plastic – we still haven't quite studied its effects. We do know the effects of microplastics on fish, and it's *terrifying*.' He is blinking in horror. 'They have altered brain chemistry, altered metabolism, altered social behaviour. We're set to have more plastics in the ocean by weight than fish by 2050. Fish are going to be like the cigarettes of our generation: doctors used to recommend them, but now we're like, "Holy shit, it's lung cancer!" Fish will go the same way when we actually study the effect of bioaccumulated plastic on human physiology.'

And then there's how you make clean fish. 'Fish cells are very robust – they grow very easily, they don't require a lot, they can deal with very wide temperature fluctuations. Land animals' cells grow at thirty-seven Celsius, whereas fish cells can do any-thing from twenty-two to twenty-six, which is way better – that's the temperature here,' he tells me, gesturing towards a window where ash fog has blotted out the California sunshine. 'The structure is a bit easier – a steak has complex, swirling marbling, whereas salmon sashimi is just layers of muscle-fat-muscle-fat, so it's easier to build. It seemed like an easier scien-tific project.'

Mike grew up in Boston, surrounded by seafood. 'I got all the Jewish stuff, like the lox, and then I got all the Boston stuff because my family was not very religious so I also had lobster and clams and crab and everything else that Jews are not supposed to have.' He went vegan after he read Peter Singer's *Animal Liberation* at fifteen, and then he met Brian, who he says is 'a genius' biochemist, at the University of Massachusetts Amherst. After a year teaching English in China, Mike went out for an Impossible Burger with Brian in New York. They had 'a few too many' beers, and decided to write a business plan.

In March 2017, they got their first investment – seed funding, a lab and co-working space from life science start-ups accelerator IndieBio – which meant they had to move to San Francisco. ('That's literally the reason why we moved to California. We had no desire to be here at all.') Now they have investors from across the globe. Among them is the venture capitalist Tim Draper, one of the first to sink money into Elizabeth Holmes's notorious blood-testing start-up Theranos, and one of the very few who stood by her even after she was charged with fraudulently deceiving investors by massively exaggerating what her technology could do.

But there doesn't seem to be any exaggeration in the Finless Foods labs. On my tour here Mike talks me through the entire process, without any jargon or dazzling theatrics, and I properly understand it this time. They get their biopsies from fish famers, university labs, sport fishermen and even San Francisco's Aquarium of the Bay. They separate the starter cells and suspend them in a solution in their 'main workhorse lab', then they filter out the cell type that is capable of expanding and dividing, and these cells are plated up into seed trays and incubated. It takes around a day for the cells to divide. 'Our European sea bass cells proliferate like crazy,' Mike says, like a proud dad. And once the cells have reached a critical mass, they go into one of the three different kinds of bioreactor they are experimenting with at the moment.

We go into Finless Foods' second lab, their 'molecular biology lab', where they grow their medium. Like JUST, they've found an animal-free serum formula, but they didn't need a Discovery Platform to discover it. 'It's salts, sugars and proteins,' Mike says, simply. 'The salts and sugars are food-grade stuff that we buy from food suppliers – nothing that people aren't already eating – and then the proteins are made from yeast. We look inside a fish and see what proteins are useful for cell growth, and we find what DNA makes those proteins. We put that DNA into a microbe system – it could be yeast, it could be something else.'

'Isn't that genetic engineering?'

'It's the same thing we do to produce rennet, which is how we curdle cheese. If people are like, "Oh my God, this is GMO technology!" it's like, "If you eat cheese, you already eat this." We're just using it to create a different protein, which is already found in fish.'

We settle into the now-empty boardroom, which has two 'Tuna of the World' charts framed on the exterior walls, but is bare and brilliant white inside.

'If you were to produce something now, it would be a kind of paste, wouldn't it?' I ask.

'Yes, that's true. We do want to find a way to use the paste as an ingredient, because it still tastes like fish. We're focusing on a spicy tuna roll, but instead of a spicy tuna roll it's a spicy *bluefin* tuna roll,' Mike effuses. 'Is that popular in the UK? Everyone eats it here. We were trying to find the hamburger of fish for Americans, and it seems to be the spicy tuna roll.'

But their big ambition is to make fillets, and they could use food science or tissue engineering to do it. 'There are a lot of different technologies that are very promising, in terms of making this look 3D; Mike tells me, whipping out his iPhone and showing me a YouTube video made by a Dutch company called Vegan Seastar, in which bearded twenty-somethings eat slivers of perfectly layered, glistening pink non-fish – 'zalmon sashimi' – garnished with sesame, from black crucibles. 'We are thinking of building something like this using food science, material science, to create a texture using plant-based proteins, or plant-based anything, or mushrooms, and then you seed it with the cells that we make as a flavouring agent.'

It sounds great if they can get it right, and I can imagine it all going horribly wrong. As I learned at JUST, food needs to look, taste, smell *and* feel right for your brain to accept it. And Mike has a greater challenge than Josh; consumers don't know what raw, unseasoned chicken is supposed to taste and feel like, but sashimi has given us very specific ideas about raw fish. Mike can't use the smoke and mirrors of cooking his product in butter or covering it with breadcrumbs. If he's making a fillet, it has to work straight out of the fridge.

Perhaps tissue engineering would be a safer bet. Finless already has a tissue engineer on staff, and Mike talks about 3D organ printing as if it's happening on every street corner in the Bay Area. 'The machinery is expensive, but what's appealing about this technology is that it's fast. You can print an organ in thirty seconds. We like it. We're looking into it. But at the moment, it's still kind of early for us.'

When Finless tuna first goes on sale, it will cost the same as an equivalent amount of conventional bluefin: around $7 for a piece of sashimi. Mike says it will be 'years, not decades' before that happens, and the critical factor is not science but regulation. I'm expecting him to launch into a well-worn diatribe about red tape getting in the way of progress, but instead he says, 'We really want to go through a regulatory system that isn't seen as skirting it. We're not a scooter company. You can't just make the technology, throw it on the street and hope for the best, because people don't forgive, in terms of food. Food is very personal. If we're seen trying to get around regulation, that's really going to bite us in the ass.' And if the first clean meat ever sold is on the market in a country chosen for its amenable standards on food safety, that will bite the entire industry in the ass. But I'm trying to forget about the JUST nugget.

The first regulatory issue clean meat has to deal with is its name. The FDA doesn't like it, and although Mike once told a reporter that clean meat is a great term because it will mean a 'clean conscience' for carnivores, it turns out he always hated it. 'It doesn't make sense in any other language. In Chinese, it sounds like you've dipped it in bleach and scrubbed it. But I was eventually convinced that the term itself doesn't matter – consistency is more important. So I changed my view and I started using it.' Mike prefers 'cell-based meats'. 'There's animal-based meats, there's plant-based meats and there's cell-based meats. It's neutral.' But it's also nonsense, because both plants and animals are also made of cells. 'It needs to be called fish, no matter what, because fish is an allergen. We need to have the word *fish* clearly on the packaging, and what type of fish it is. But I do want there to be a firm delineation, because what we make is better. There are so many advantages to what we're doing, I want people to buy it on purpose.'

Mike is sure that one day the meat being grown in labs, what-ever it is called, will replace conventional meat. 'First it's going to be small, it's going to be an ingredient, it's going to be part of a plant-based product, a hybrid product, and eventually it will be what people really want it to be. People think scientists have a lot more figured out that we actually do – we do not.'

I think about the bit in the JUST chicken video when Josh boasts, 'We've figured out how life really works,' and realize what a breath of fresh air Mike is in this Silicon Valley ash cloud, and how there might be some real substance to this industry, after all, if there are more scientists like him in it.

'There's a lot of hype,' he continues. 'It's going to be slower and smaller at first than people think. But it's definitely going to happen. I'm not saying Finless Foods is inevitable, but the technology is inevitable, and this is how people are going to eat, unless we wipe ourselves out first.'

'I'm going to make a terrible pun,' I say. 'Are you a fish out of water, in this Silicon Valley start-up world? Do you fit in here?'

'I fucking hate it,' he says. 'We've been trying to leave since the second we got here. The culture here is weird. Sometimes we're in these meetings with people and it just feels like they're aliens. Hopefully we can be somewhere else, eventually.'

But his foreignness goes beyond being from the other side of the US. As well as being a CEO, he tells me he is also a communist. 'I wouldn't say that a lot of our investors love communism,' he smiles.

'Do *you* love communism?' I ask. 'Are you properly a communist?'

'I would say so, yeah.'

'How can you be a communist entrepreneur?'

'I'm trying to build a technology that I think is important. I'm trying to do something that I hope will shape the way that we eat, for the better. The current mechanism to do that is a start-up. I wish there was a different system. I wish we had a better way to do this, but currently we don't.'

'Do you really not care about making money?'

'In order to make sure the relationship with our investors stays good, it does need to be a profitable business. For me, personally? Not really. I already make more than I need. I give a lot of it away. I make, like, $85k, which is fine. I'm not married, don't have kids. My co-founder and I make among the lowest salaries in the company.'

'Transparency is supposed to be so important in the clean meat world, but when I actually try to talk to people there's not a lot of it,' I say. 'How come you are so happy to tell me all this?'

'I think a lot of our story is the fact that we are more genuine. That's how this will win,' he replies. 'Look at trends in terms of what millennials and Gen Z are interested in: we get revolted by bullshit. Anything that seems even a little bit corporate, a little bit polished, we reject. We don't look corporate here. We are trying to be genuine, and it's our brand.'

In other words, Mike's openness is a deliberate branding exercise, another way to differentiate the company from other start-ups in order to 'win' the market as the millennial meat makers.

But there's one thing that remains profoundly opaque about Mike Selden: his veganism. Even though he speaks in the language of animal rights and has effused about his veganism in every previous interview I've managed to find with him, today he tells me he is not vegan anymore. 'I buy entirely vegan groceries. I mostly eat at vegetarian and vegan restaurants. But I just don't call myself vegan in part because I don't want to be nitpicked, now that I am a mildly public figure.'

And then he tells me a story about how he was speaking at a conference recently, and a woman approached him afterwards to ask what app he used to pick his wine. Mike told her he didn't use any app, so she told him that meant he couldn't be vegan, and he said, 'OK, then: I'm not vegan.'

'The vegan community is like the most self-absorbed group of people who are so unable to see outside of their own thing.

It's incredibly white, it's incredibly wealthy, incredibly privileged, and it's super unaware of what it does. I just didn't want to be associated with that,' he says.

Mike clearly is vegan, but he knows how impossible it is to be a perfect vegan, and he doesn't want to be accused of being a bad vegan, so he'd rather say he isn't vegan. I feel sorry for him, grateful that I have never claimed to be more than a heartless carnivore, and glad to be too old to be part of Gen Z, where any perceived transgression can leave you outcast. You'd have to be a contortionist to live a pure enough life.

But however he defines himself, Mike thinks veganism is going to be obsolete once this inevitable technology gets worked out. 'We don't want it to be *seen* as vegan, we want this to be *food*. My hope is to make everybody vegan without changing their habits.'

*

Hardcore vegans don't do things by halves. In 2004, English animal rights extremists targeted a family-run Staffordshire farm that bred guinea pigs for scientific research. They sent fake bombs to their cleaner, leafleted the neighbours of the man who delivered their fuel defaming him as a convicted paedophile, and spelled out the name of a farm labourer in shotgun cartridges outside his home. When that wasn't enough to close the farm, they dug up the corpse of Gladys Hammond, the deceased mother-in-law of one of the brothers who owned it, and left messages saying her remains would only be returned once the farm was shut down. Three activists were eventually sentenced to twelve years in prison each.

Animal rights campaigners have mellowed in recent years, but only a bit. A month before my visit to Mike, Whole Foods took out a restraining order against Berkeley-based vegan activists Direct Action Everywhere (DxE), who were planning to protest the welfare conditions of Whole Foods chickens in their local branch, ten minutes away from Finless Foods. DxE had previously acted out scenes of animal slaughter in the meat and dairy aisles, and splattered eggs with fake blood. Elsewhere in Berkeley, DxE activists lay naked, soaked in yet more fake blood and wrapped in plastic outside a family-run butcher every week for months, accompanied by sound recordings of pigs squealing in terror, until the owners agreed to put a sign in the window that read, *ATTENTION: Animals' lives are their right. Killing them is violent and unjust, no matter how it's done.*

So I was expecting there to be some kind of backlash from more militant vegans against the clean meat industry; after all, clean meat actively encourages people not to change their eating habits, and to continue living at the expense of animals, albeit the greatly reduced number required to provide starter cells. Accepting clean meat would mean condoning a technology developed using animal experimentation and FBS, and buying it will line the pockets of big meat companies, like Tyson and Cargill, who have invested heavily in the clean meat start-ups and have been responsible for the slaughter of billions of animals worldwide. I thought at the very least there'd be some kind of online campaign, perhaps a bit of chanting and protesting in the Bay Area, maybe even some entrepreneurs being drenched with mock foetal bovine serum on the way home from the lab.

But there has been barely a squawk from the vegan community. When Mark Post presented his burger to the world in 2013 there were a few rumblings that the idea was a bit gross; the Dutch Vegan Society put out a poster campaign pointing out how much more appealing a veggie burger looks compared to a slab of flesh in a Pyrex flask. And that has pretty much been it, in terms of organized opposition to clean meat. I call the Vegan Society in the UK, and the press officer says they think clean meat is 'very exciting'. I ring up Wayne Hsiung, DxE's co-founder, to see what their take is on the industry blossoming in their back yard, and he tells me it's 'part of the solution' to animal exploitation. 'As long as it doesn't obscure the consequences of using animals,' he says nebulously, 'it's going to be beneficial.' Militant vegan YouTubers are cautiously optimistic about the technology. I scour the normally uncompromising comments underneath their vlogs on clean meat. Nothing.

It is only when I plunge the deepest depths of Google that I find a 2010 paper written by an outlier, a lone dissenting vegan voice: a British sociologist called Dr Matthew Cole. 'IVM [in vitro meat] ignores the powerful vested interests and social forces that create "demand" for meat and that routinely stigmatize veganism,' it reads. 'In fact IVM further stimulates "demand" for meat by perpetuating a myth that meat is and will always be intrinsically desirable.'

These words were written years before clean meat start-ups ever existed, but they say something very prescient, because the entire clean meat industry is built on the premise that the desire for meat is natural.

At JUST, Josh Tetrick had said to me, 'I miss meat. I love meat. I want to move towards it and smell it and look at it.' When he tried JUST chicken for the first time, he told me, 'In a primal way, I was experiencing the thing that I really miss.'

'Do you think it's a primal thing?' I'd asked. 'That we're primed to enjoy meat?'

'I do think there's an element of it. Human beings have been using spears to kill animals for thousands of years and building symbols and artefacts and cultures and community around ideas around that. You can ignore it or you can go with it.'

But could the belief that our taste for meat is hard-wired really be nothing more than a myth?

I meet Matthew Cole at the headquarters of the Open University in Milton Keynes – a modernist, grey campus, empty of students; a kind of academic ghost town. Matthew is waiting for me at reception. He's short and slim, bald, with smile lines. We head to the cafe to pick up a coffee from one of those fancy self-service machines, and I'm just about to ask him where the milk is, before I stop myself and decide to have it black.

Matthew is a vegan sociologist, in every sense. His work specializes in the sociology of human–animal relations, how children are socialized into accepting human domination of animals, and how vegans are represented in the media. He's shot a few videos for the Open University's YouTube channel. There's one entitled *Dr Who Should Be Vegan*. 'Love for life in all its forms is one of the central messages of *Dr Who* and a great reason for its popularity,' he says, unsmiling, looking directly into the lens. 'The time is long overdue for a fully morally consistent and vegan Doctor.' The top-rated comment reads, 'This guy looks like he could do with a steak.'

'You wrote something in 2010 on in vitro meat. Do you still call it that?' I ask.

'I do.'

'Why?'

'Because it sounds bad,' he grins. 'The terminology that we use to describe in vitro meat, cultured meat, whatever it might be, is part of a discursive game, or battle, or war, if you like, to construct the meaning of what this substance is. From my point of view, it's a bad thing.'

Matthew is worried about 'the class dimension' of meat grown in labs: that it will be sold as an elite product that will lead to a moral hierarchy, where the affluent people who can afford it will be able to reinforce their superiority over the people and nations that can't. 'Here's the rational white man going around the world saying, "Our methods are superior to your barbaric ways,"' he explains. It also stops us questioning the human drive to subjugate everything around us. 'Nothing needs to change for in vitro meat. That's why it has this appeal: everything else could remain unchanged. It will not change the fundamental relationship of human beings to animals, to the environment, to the natural world; it would still be a relation of domination.'

'Why hasn't there been a big backlash against it from vegans?'

'It's alluring. The superficial promise of it is to eliminate 99 per cent of animal agriculture – obviously, I can see that is exciting. And I suspect a lot of activists see that as a quick win. We've put in all these decades and decades of effort, we don't seem to be getting as far as we'd like nowhere near quick enough; maybe this could short-circuit that struggle.'

Matthew has written papers on what he calls 'vegaphobia': the stigmatization of veganism and vegans. It's interesting to me, now that I've met so many people who want to keep their veganism under wraps. Matthew has put the negative vegan stereotypes that circulate in the mass media into five distinct categories: 'Vegans can be portrayed as hostile, soppy, wishy-washy, just following a trend, or just flat-out ridiculed.'

'Is this something you have experienced yourself?'

'Yeah. Especially through doing my academic work, in terms of doing public things – YouTube videos or articles for The Conversation. You just have to look at the comment threads. There was a paper that I wrote with Kate Stewart, who is my partner as well as my colleague, about the film *Sausage Party*. I don't know if you are aware of it?'

It's an R-rated spoof Pixar film about a talking sausage called Frank and his girlfriend, a talking hot dog bun.

'Sounds great,' I say.

'I cannot recommend it,' he replies, gravely. 'We wrote a paper critiquing it – a vegan critique of this film. And it was picked up by a Twitter account that tries to lampoon academics. They trawl for articles that look stupid and say, "Isn't this funny? Ha ha ha."'

I don't want to stereotype Matthew as a hostile vegan, but he definitely doesn't see the humour in any of this.

'Vegans are aware of those kinds of negative stereotypes being out there,' he continues, 'and sometimes there is anxiety about being seen to reproduce them.'

'Why do these stereotypes exist?'

'There are loads of vested interests behind animal exploitation. They are huge and immensely powerful, with a long history behind them. A huge amount of cultural labour has gone into reproducing, legitimizing and defending animal exploitation in popular culture, but that's underwritten, supported, by state activity – by nutritional education. It's all interconnected. It's massive. And sometimes it feels impossible to defeat it.'

But surely the desire for meat goes beyond those vested interests. We are hunter-gatherers, after all. It's human nature to kill animals and eat meat. 'Haven't we evolved to like the taste of meat? Isn't it natural?'

'No. Humans are highly adaptable creatures, inventive and creative. We have transcended biological and environmental limits in many ways.' He gestures to the sleet falling behind the window. 'You could argue we shouldn't be living here – it's too cold for the human organism. The same is true of our consumption of animal products. Nothing about that is natural.'

'So where does our desire for meat come from?'

'It's a cultural construct. The availability of animal products is self-evidently an outcome of social processes. It is not natural. There could never be anywhere near enough edible non-human animals on this planet to sustain the current level of human consumption of them without artificial intervention. And drinking the milk of another species is completely bizarre. There's nothing natural about that, whatsoever.'

And I think of my one-year-old daughter, who I last saw, smiling, with a beaker of cows' milk in her hand as I waved her goodbye this morning, and something that seemed the most natural thing in the world is suddenly disturbing.

'Eating meat is literally rammed down our throats before we can even speak,' Matthew goes on. 'We feed it to our children, and we reward children for eating meat. Before you can speak, the idea is given to you that this is the tasty thing. The message is very powerful – it comes from your mother.'

I know from my own experience that Matthew is right. Milk, eggs, cheese, fish and meat are promoted in government campaigns and parenting books as essential foods to give to your children. When I first became a mother, I went on a free weaning workshop run by my local council. The message there was that parents shouldn't leave it too late to introduce meat, and that a vegetarian diet wasn't healthy for babies because they need iron for their brain to develop properly, and it was almost impossible to get the right amounts from anything apart from red meat. So I stuffed both my children full of bolognaise before they even had proper teeth to chew it.

Matthew says it's well established that vegan diets are nutritionally adequate for babies, children and adults alike.

'If that message is wrong, why was the council giving it to me?' I ask.

'It's down to the sheer weight of cultural labour that's gone into establishing animal products as essential, as natural. For many people, still, it seems unthinkable that you would deviate from that. It does look like deviance, from that perspective. You are a deviant for not feeding your child meat.'

That night, as my daughter is enthusiastically scoffing the shepherd's pie I am spooning into her mouth, I think about how I have rammed the taste for animals down her throat, and I feel a shiver of disgust. Surely it is that feeling that needs to be harnessed and cultivated if we want to solve the problems caused by animal agriculture, not the emerging technology to grow meat in labs.

But for now it is only a *shiver* of disgust. I wipe my daughter's chin and fetch her milk.

CHAPTER EIGHT
Aftertaste

Oron Catts has made a career out of cultivating disgust. Today, he is growing mouse scar tissue in foetal bovine serum, using an incubator made of manure. 'It's sixty-five degrees Celsius, that compost pile,' he declares, gesturing towards a wrought iron cage containing a tissue culture flask on top of an imposing heap. 'It's made from woodchip and horse shit from the mounted police.'

We are standing in a courtyard of King's College, London, with the Shard so close it's almost impossible to see its pinnacle, beside a truncated pyramid of dung. This is Oron's latest artwork, entitled *Vessels of Care & Control: Compostcubator 2.0*, and he has travelled from the University of Western Australia in Perth to see it exhibited. The strangely beautiful compost heap is the first piece visitors see as they enter the Science Gallery London's *Spare Parts* exhibition. The Compostcubator uses principles of permaculture, with microbes in the compost generating the heat necessary to grow mouse connective tissue entirely off-grid. It's supposed to make us examine how human beings think we can control and replicate life. 'It will be one of the first times that a piece of cultured mouse will be presented outdoors,' Oron says with pride.

Oron has spent twenty-five years using living tissue as his medium of artistic expression. Along with his partner in art and life, Ionat Zurr, he has grown wing-shaped objects out of pig tissue (*Pig Wings*, 2000), a living jacket made from cultured mouse cells (*Victimless Leather*, 2004) and made a domestic bioreactor for farming in vitro insect meat (*Stir Fly*, 2016). But he is also perhaps the most unsung pioneer and unwitting trailblazer in the world of clean meat. In 2003, his *Disembodied Cuisine* exhibit was the first time on earth anyone had ever grown and eaten in vitro meat, nearly a decade before Mark Post lifted the cloche on the Sergey Brin-funded burger. With a single five-gram frog steak marinated in Calvados, Oron jump-started the industry that is exploding in Silicon Valley and beyond. He's now its most outspoken critic.

Hardly anyone in Silicon Valley knows his name, but Oron is a distinctly memorable man. He looks like a wizard: his mesmerizing beard is long, curly, bushy, grey and very pointy, and his hair is slicked back into a ponytail of tumbling curls. He has so much to say, and he says it too fast. I wanted to meet him to hear about the frog meat, but when we sit down to talk he wants to tell me the entire story of his professional life. My questions almost get in the way.

'My background is in product design,' he begins. 'What I recognized in the early nineties – and what is becoming painfully obvious now – is that biology becomes an engineering pursuit, and life becomes a raw material to be engineered. It provides a new palette for artistic possibilities.' Instead of designing biological products, Oron chose to become an artist. 'I felt that as an artist I have a licence to problematize things rather than be a solutionist.' In other words, Oron is allowed to ask questions without any obligation to answer them.

He calls his creations 'contestable objects'. 'I found the whole idea of designing with life a contestable idea, not something we should accept at face value.'

'Although a lot of people do,' I manage to interject.

'Certainly, that's right, and it's becoming worse and worse. And in a place like San Francisco, you realize those people have no trace of self-reflection.'

Meat has been on Oron's mind since he grew up force-feeding geese for foie gras on a farm in Israel. In the mid-nineties he teamed up with Ionat, a scientist, who taught him the techniques of tissue culture. 'It's not difficult to learn how to do. It's a craft, not a science,' he says, tugging on his fabulous beard. 'I thought I might be on to something that could solve problems in the world. The more I dug into it, the more I realized it's an extremely problematic approach.'

Oron says humans are not ready to control biological systems because we don't properly understand what life is. If the cells in the corneas of a rabbit are still alive hours after the heart has stopped beating, is the rabbit still alive? Is it semi-alive? 'We only have one word for *life* in the English language, whereas we have fifty words to describe *shit*. So we can't even put into words what we are doing.' And that mindset, that lack of nuanced understanding while tinkering with life, could lead to ultimately horrific possibilities. 'We suffer from cultural amnesia when it comes to our control of living systems. What we choose to do with life, we will end up doing to ourselves.' The systematic breeding of animals led to the eugenics of the twentieth century, he says; who knows where the systematic growing of animal flesh will lead us.

'The solution to the problem that in vitro meat is trying to solve can be much more easily solved by the reduction of meat consumption. From a perspective of efficiency, it's overshoot engineering,' he tells me. 'But it produces this seductive narrative that everything will be OK, we don't need to change our behaviour, because those smart scientists will figure out a way, business as usual, and we can increase consumption.'

The *Disembodied Cuisine* installation, which took place in a converted biscuit factory in Nantes, France, in March 2003, was always intended to provoke uncomfortable feelings. 'We played on what constitutes foul food. We knew that French people don't like the idea of engineered food much, and we chose frogs because most other cultures find the idea of frogs unappetizing.'

They built a tissue culture lab and a dining room in the gallery, behind plastic sheeting curtains emblazoned with biohazard signs. They cultured cells from the African clawed frog for three months, while members of the public looked on. On the final day of the exhibition, six people – Oron, the exhibition curator, the director of the museum and three members of the public – ate the frog meat. (Ionat was pregnant and excused from the table.)

Oron flips open his laptop to show me a film of the history-making frog-eating culmination of the artwork. The diners sit at an immaculately laid table. Oron is dressed as a waiter, but wears latex gloves. He still has the beard, but it is shorter and blacker. A French chef fries the Calvados-marinated frog steaks in a miniature pan on a camping stove, and the diners smoke while they wait to be served: very artsy, very French, very much a product of another era. Then the globules of frog are dished up with tweezers onto large white plates. *'Bon appétit!'* someone says, and the diners slice into the meat with scalpels. No one looks like they are aware that they are about to make history when they put it into their mouths.

'I was quite concerned about health and safety, so I asked the chef to cook it in a garlic and honey sauce, which are well-known antibacterial agents. The sauce was amazing,' Oron remembers.

'We were able to grow about five grams and to distribute it be-tween six people. It was the ultimate nouvelle cuisine.'

But there was a problem with the polymer scaffold on which the frog tissue had been grown. 'The polymers were designed to break down within the context of growing mammalian cells and warm-blooded cells, at thirty-seven degrees. The frog cells were growing at room temperature, so it didn't break down properly. The polymer is like felt, so it still had a very strong texture of fabric, and the frog cells, even though they were muscle cells, we didn't exercise them. They were more –' he searches for the right word – 'jelly.'

'It sounds absolutely disgusting.'

'Exactly!' he exclaims with delight. 'Three of us swallowed it, three just couldn't. They spat it out, which was great for us because we could use what they spat out in the follow-up exhib-ition, which was called *The Remains of Disembodied Cuisine*.'

It is all so arch, so knowingly playful, and it's a missed oppor-tunity, in a way: Oron's critique is presented as a beard-stroking curiosity, something for a tiny audience of art lovers and intel-lectuals to ponder, rather than the catalyst for a very necessary and wide public debate about the future of food. The first piece of clean meat ever eaten was produced in order to highlight how problematic this potential technology could be, and the world received the product but not the message alongside it.

'We anticipated that it would generate interest, but there was very little coverage,' he concedes. 'The main thing was that the world was busy with something very different, and that was the second Gulf War in Iraq.'

Oron and Ionat moved on to other things, like the tiny jacket made of living mouse tissue (which the curator at the Museum of Modern Art in New York had to 'kill' by switching off the incubator because it was growing too fast). Meat wasn't their focus, but it was pretty much off the menu: Oron says he gave up eating anything warm-blooded after the frog. Then, in 2011, someone sent him a link to a story about a Dutch scientist who was claiming he was going to be the first person to grow and eat in vitro meat, and was planning to do it in some kind of live show. 'This was kind of amazing. This was too much.'

The Dutch scientist was Mark Post, of course. Oron sought him out and got him to agree to participate in another of his artworks, 2012's *ArtMeatFlesh 1*: a cooking show in Rotterdam in front of a live audience with judges and tastings, and a debate between scientists, artists and philosophers about meat. None of the meat cooked was lab-grown, but every dish contained something disgusting and thought-provoking that could be the future of food, be it mealworms or FBS. 'It was a real multi-media experience, and very enjoyable for everyone. We were able to get into some very serious conversations,' Oron says. 'Mark played along, that's why I have so much respect for him. And he likes to cook. He was wearing a chef hat.'

There are clips of *ArtMeatFlesh 1* online. There's Mark, the esteemed scientist and father of clean meat, complete with chef's hat, laughing and joking and dishing up revolting things. Although it's in many ways the opposite of his humourless 2013 burger reveal, when you watch the two events alongside each other it's obvious that Mark borrowed several ideas about how to engage an audience – how to put on a show – from being part of Oron's piece. There's a heavy irony to all this: Oron's work was only ever meant to be a performance, and now we have a clean meat industry based on performance, from Mark's burger to the JUST nugget.

'You're the first person to have grown meat in a lab and eaten it, but nobody knows it's you. How do you feel about that?' I ask.

For the first time in an hour of talking, Oron pauses. 'I have an ego. I care about it to some extent,' he finally says. 'One thing that I found amazing – this is how fucked up the media is – after Mark's burger thing, only two media outlets in the whole fucking world approached me for comment. One of them was *Time* magazine, and one was a rural ABC radio show. I spent quite a lot of time with the *Time* reporter, telling her the whole story, and it ended up being just one tiny sentence. She emailed me an apology and said, "Unfortunately the editor didn't feel that your story contributed to the narrative that we wanted to have." They wanted to have a good news story.' For a moment there is a bitterness in his voice. But then he adds, gently, 'Mark is interesting, because there are a few cases where he actually did give us credit, in a kind of offhand way.' But the world-saving burger in a Petri dish is a much neater story than the vomit-inducing frog meat nouvelle cuisine, so that's the story that gets told.

'Your *ArtMeatFlesh* thing and Mark's launch have a lot in common. Maybe his burger wouldn't have had as much impact if it wasn't launched as a performance?'

'This is where it makes things quite powerful. This is one beautiful example where science follows art.'

'But how do you feel about being the unintended forefather of this new industry, an industry that clearly troubles you?'

'It's not what we intended, but quite an important part of our work is our critique of the psychopathologies of control: we're trying to control systems that existed outside of our control for millennia,' he replies. 'One thing which was very important for us from the beginning of our practice was to not try to take control of it. Once our work is out there in the public domain, it would generate its own stories and narrative.' He smiles. 'I'm fascinated to see where it goes.'

*

There is no campaign against clean meat. The few individual voices I've found who are prepared to criticize it are over-whelmingly drowned out by the chorus of positive messages sent out by the clean meat industry. But, despite the culture of irrepressible inevitability promoted by the start-ups and the GFI, no one has any idea where clean meat is going to go.

Bruce, Josh and Mike had been so confident that consumers were going to accept clean meat, that they wouldn't care that it came from a lab, that they would prefer it to meat grown in animal bodies, but the 'ick factor' is actually a serious problem for the industry. Bruce is totally unfazed by any suggestion that people are grossed out by the idea. 'I'm not concerned when polls show some portion of the population is no more eager to accept in vitro meat than their grandparents were to accept in vitro babies,' he wrote in the *L.A. Times* in 2018. 'There will always be Luddites who decry and resist new technologies. That's to be expected. But the rest of us will happily enjoy conscience-clearing clean meat.'

Yet the conscience-clearing benefits of clean meat are also contested, as I discover once I take a proper look at the few academic papers available that examine the claims made by the industry and GFI. Most worryingly, I found at least four pieces of research that conclude that, while it might be more efficient in terms of land, water and energy use than beef pro-duction, clean meat produces more greenhouse gases than raising poultry – as much as 38 per cent more, according to one study. We'd be better off eating chicken to save the planet. (In fact, two of these papers say we'd be much better off eating insects, but that's another challenge, as far as ick factor goes.)

All of these studies used very speculative estimates of the inputs of clean meat production; scientists and entrepreneurs are still working out how to grow meat in labs, and production methods are bound to grow more efficient. But the point is that nobody can really say for sure whether clean meat is better for the planet at the moment, and that's a worrying ambiguity given the certainty with which its conscience-clearing environmental benefits are being sold to investors and consumers today.

And, of course, clean meat is still bad for us. The risks of eating mountains of red meat won't go away just because it's been grown in a lab. It will still give us cancer and heart disease, it still has cholesterol, fat and no fibre, even if it can be engineered to be a little better for us one day. The danger is that, if we are told the meat we are eating is 'clean', we might feel that we have a licence to eat as much of it as we please, and it is nonetheless more damaging for the planet and for our bodies than a plant-based diet.

So is plant-based meat the answer? Those bleeding Impossible Burgers and juicy Beyond Burgers? Maybe. Perhaps not. Plant-based imitation animal products are ultra-processed foods, made from an eye-watering number of components. When I look up the ingredients of the JUST Egg I ate it reads like the apparatus list of a chemistry experiment, a roll call of isolates and gums and oils and extracts and flavourings, tetrasodium pyrophosphate, transglutaminase, potassium citrate and more. The Beyond Burger is billed as being made of pea protein and coconut oil, but it also contains something called methylcellulose, maltodextrin, vegetable glycerine, gum arabic and succinic acid. You need to do a lot of tinkering to turn plants into something resembling animal products. And when you add up the miles required to ship all these components to the factory, and the nutrition they all provide, or don't provide, in comparison to vegetable dishes that aren't pretending to be meat, that anyone could make from ingredients they can grow in their back garden, it seems like quite a silly idea to be going to all this effort.

Vegan meat depends on a pessimistic view of human beings: the belief that we are incapable of changing the way we eat. But the only way to be absolutely sure our food isn't costing the earth is for us to lose our taste for meat. After all, the problem isn't really animal agriculture, it's human appetites.

This doesn't have to be about absolutes, though. 'Even the possibility that this technology slows future potential increase in livestock meat would be a form of victory, a form of success,' says Dr Neil Stephens, a sociologist at Brunel University, who probably knows more about the industry than any other academic in the world, and is the only person I've spoken to so far who is at pains to be even-handed and cautious. Neil is vegan, but it feels incidental to his work. He has been studying clean meat since 2008, looking at the politics, ethics and regulatory issues this form of food production would create, and I've just read a paper he wrote on 'challenges in cellular agriculture' which is so balanced it almost knocked me off my chair. I've rung him up in search of some much-needed sanity.

'If the clean meat industry gets it right, and works out how to make something that's really equivalent to meat, what challenges should we be concerned about?' I ask.

'Concern is too strong a word,' Neil says, carefully. 'We should be *mindful* of what the implications might be. Currently the technology is being developed by sets of companies and people in universities supported by a whole other set of people who are all genuinely concerned by the state of the world today, and genuinely committed to dedicating their lives, their intelligence and their passion to doing the best thing to address that through technology. You would expect, looking at other start-up cultures, that ownership could well change, through licensing, or companies being bought out. Who may be owning this technology in twenty years' time, and what their values are, and how they relate to profit margin, may shape how it's used.'

This is a potentially enormous concern, no matter how tempered Neil's answer might be. We can't control the direction of the market. We can't control who will run the clean meat industry of the future, and it might not be well-meaning vegans, it might not be Mike the nerd or Bruce the evangelist. It might be someone with very different priorities indeed.

'If it works, you could imagine having a commercially successful sector, companies that make money that don't have anything like the significant social and environmental impact that is suggested if it remained small scale,' Neil continues.

I think about all those big investments from giant meat companies that the start-ups are so eager to secure, corporations that are notorious for putting profit above the welfare of animals, people and the planet.

'Would it be the companies that already have access to the infrastructure and logistics clean meat needs that would be taking it over?' I ask.

'That seems a quite possible, maybe even likely scenario,' he replies.

For all Bruce's idealism and Mike's communism, they might be helping existing meat companies get richer, and doing the groundwork for an industry that makes us all reliant on ever more remote multinational corporations. In the future the clean meat industry is fighting for – where humans still eat meat, but no longer kill animals – we will have surrendered our self-sufficiency to companies with specialized technology. No one can guarantee that these companies would be a force for good, or run for the benefit of anyone other than themselves.

*

To understand where something is going to end up, sometimes you need to return to the beginning. After months of emails, I'm finally sitting opposite Mark Post. And he is telling me how often he eats sausages.

'Every day, to be honest, because I put slices of sausage in my sandwich in the afternoon,' he says, every inch the Dutchman despite his American-tinged accent. 'And in the evening we sometimes eat meat too. I eat meat as much as anyone.'

I've come to see Mark at the University of Maastricht, where his crumpled brown shirt and dark green trousers clash beautifully with his office's orange carpet and yellow walls. He is even taller than Mike Selden, with a small paunch, receding grey hair and a hearty chuckle that peppers our conversation – 'a ha ha ha ha' – like machine gun fire. Mark is Professor of Physiology here, but he's also a cardiovascular surgeon, the chief scientific officer of Europe's biggest clean meat start-up, Mosa Meat, and a very busy man. I'm lucky to be here. But Mark is lucky too, because, the way he tells it, the entire cultured meat industry only exists because of a series of accidents, absences, coincidences and unintended events.

It began because of the passion and determination of an eighty-one-year-old man, Mark explains. Willem Van Eelen was a Dutch entrepreneur who had dreamed of the idea of victimless meat cultured from cells ever since he experienced brutality and starvation as a prisoner in a Japanese POW camp. Van Eelen knew he had to hustle to make his dream a reality. 'He coerced scientists from three universities, Utrecht, Amsterdam and Eindhoven, to submit a grant to the Dutch government to make this happen,' Mark tells me. The Dutch government agreed to stump up enough money to fund a cultured meat project for five years, starting in 2004.

But there was limited enthusiasm for it. 'None of the scientists initially involved were actually interested in making cultured meat. They were all using this as an umbrella to do their thing.' They worked on the project in as far as it could advance their existing research interests. Eindhoven, for example, were much more dedicated to making a model system for bedsores than anything edible. Mark came on board two years into the project, after the project leader in Eindhoven fell ill. 'I just thought it was a great idea. The more I learned about it, the more excited I became.'

Mark's eyes sparkle when he talks about his work. His contagious enthusiasm has been crucial to the success of clean meat, but his communication skills were only revealed because of another set of coincidences, in 2009. 'I was on the train recovering from a boring meeting in The Hague one rainy Thursday (most meetings in The Hague are very boring, a ha ha ha ha), and I got a call from a journalist from the *Sunday Times*. I actually didn't quite realize what the *Sunday Times* was.' Neither of the academics who dealt with press enquiries about the project were available, the journalist said – could Mark answer a few questions? 'I had nothing else useful to do, so I said OK. And that was the beginning of a media frenzy, because she put it on the front page, and AP and Reuters sent it all over the world. All of a sudden, I was the point person.'

After the money from the government ran out that year (the Dutch Ministry of Economic Affairs didn't see any commercial potential in what was being produced – 'I know they regret that now,' Mark chuckles), he was well versed in the power of media coverage, and the momentum it might be able to give the project's funding. And he had seen from Oron that you could turn making meat into an entertaining show. 'I thought, Why don't we make a sausage, present it to the public, and have the pig that donated the cells for the sausage honk around on the stage,' Mark says. The pig would be a living advert for the research they were pioneering.

But even that sausage would take €300,000 of ingredients and labour to make. Mark was plodding on with limited funds when he got a call out of the blue from what later turned out to be Sergey Brin's office. 'They wanted to talk to me about what I was doing, and I said, "Sure." At that time I talked to *everybody* about this project, so why not.' One of Brin's right-hand men flew over to Maastricht on a Dutch public holiday, and Mark told him about his plans for the pig/sausage performance.

'And then he said, "Well, Sergey wants to fund this." I had *no idea* who Sergey was. He said it as if everybody should know him, so I thought I should pretend that I did. A ha ha ha ha.'

Mark had two weeks to write a two-page proposal. 'I said, "How much money should I think about asking for?" He said, "Oh, a couple of million." I said, "We can do that." And he said, "By the way, it cannot be a sausage, it has to be a hamburger." Having no clue that it would be much more difficult to make, I said, "Yeah, OK."'

'Why did it have to be a hamburger?' I ask.

'Because it's America.'

'Why is a hamburger more difficult?'

'Because it actually has to look like *meat*. A sausage can be anything. You can get away with anything in a sausage. In a hamburger, you cannot: you have to make fibres that appear like meat. But in the end we made it happen.'

It's impossible not to like Mark. Of all the people in the strange world of clean meat, he has the greatest claim to be taken seriously, yet he is also the most humble and self-effacing, the only person prepared to really laugh at himself. Perhaps it's because he sees how contingent his success has been. Perhaps it's because he's been an academic for nearly forty years and doesn't need anyone else's validation. Or perhaps it's because he's not in a Silicon Valley start-up.

The launch itself was at the TV studio in West London where *TFI Friday* used to be filmed. Brin's office hired the PR company Ogilvy to manage it. 'I never got the bill for that and I'm sure it was even more expensive than the entire hamburger,' Mark says. 'We actually thought of having Ferran Adrià cook the hamburger, and getting Leonardo Di Caprio and Natalie Portman to taste it. Ha ha ha ha ha ha ha!' In the end, they went for something only slightly less glamorous, to keep the focus on the science. But it was still a show, and a huge hit, while Oron's pioneering performance, which was so much more entertaining but happened without the support of a PR company, sank without a trace.

'Were you surprised by how much news it made?' I ask.

'I was, yes. I was aware of the power of the story, but I was sitting there with curled toes thinking, I hope they're not going to trash this.' He lowers his voice, conspiratorially. 'To give you an idea of how naive we all were at that time, literally on the Sunday morning before the show, Ogilvy had us all in the room and asked me, "Why are you doing this?" and I was like, "What?" I really had not thought about the message. I had to actually think, Why *am* I doing this? We came up with two reasons: one is that we wanted to show the public this was actually doable, the technology is there, and the second message was that we need to think about how we are going to produce meat in the future, that current meat production is not sustainable.

The big third thing was that we wanted to have money, but that was not part of the message. A ha ha ha.'

So the potential of lab-grown meat to save the planet was a bolt-on, an afterthought dreamed up the day before the launch at the behest of a PR company.

It was coincidence that they launched the burger on a slow news day in August, when there was no Gulf War to compete with for airtime. But the location was a strategic choice; they couldn't get the burger into America because of import restrictions. 'The only way we could have done it in the US was to do it in the Dutch Embassy, ha ha ha ha, which is of course *not* a good location. We either had to do it in the Netherlands or in a country where we could smuggle it in. Since there's a train connection with London, we could do that.'

The impact of the launch still surprises Mark to this day. 'I meet people who tell me, "We have this investment fund and it basically exists because of it." "We started this company." Or students who started studying bioengineering because of it. In retrospect, this was a very, very lucky choice.'

Mark's company, Mosa Meat, was founded in 2015 (Mosa is the Latin name for the river than runs through Maastricht). The Mosa burger will be manufactured in a Dutch factory and is expected to go on sale by 2021, initially costing nine euros.

'Do you have a plan to move to cuts of meat?' I ask.

'Yes. Of course.'

'How far off will that be?'

'Ooooh.' He takes a little Dutch speculaas biscuit from the saucer of his coffee cup. 'That's very difficult to answer, to be honest. We're starting to work gradually on that now.' He crunches the biscuit slowly. 'The theoretical framework is there. We know what we all have to do to make that happen. When it will be, to the eye and to the mouth and to the nose, a ribeye that you cannot distinguish from a ribeye that you get from a cow, that's difficult to guess. So I'm not going there.'

I think back to how nonchalant Vitor had been when he told me JUST could 'grow a steak in a week if we wanted to'. It reminds me of something else I've been wanting to check. 'Can you really take a biopsy from a feather and make meat out of it?'

'Oh, God. Theoretically, you can. It's the stupidest idea I've ever heard, to be honest. If you're going to do chicken or fish, the obvious cell source is a fertilized egg; it's the ideal source for these cells. Unfortunately, with cows, we can't do that.' But piercing an egg and syringing cells out of it wouldn't look as good in a promotional video as plucking a feather from a green pasture. 'It's possible, but it's the worst idea ever, because it's contaminated, it's been on the floor. So you have to throw a lot of antibiotics at it. And to use it internally you'll have to genetically modify the cells,' he continues. 'Actually I talked to scientists at JUST at my conference a year ago, and I asked them, "Really? What were you guys thinking?" and they said, "Well, it's not our idea. It's from marketing." A ha ha ha ha.' Mark laughs so much at this that I can see all his teeth.

But Mark says he is 'exceptionally glad' he is no longer the only scientist trying to grow meat from cells, whatever cells they might be. He is grateful for the community the industry provides. At one point he was eager to share information on things that *didn't* work so that others didn't have to repeat his mistakes, but his investors weren't keen, so collaboration between companies is limited to regulatory things, he says. His long-term plan is to develop the IP and then license it out, so his investors are happy but his technique can be a widespread, global method. And, of course, that would mean that anyone could use it, so long as they paid for it.

'There's a race to be the first to put this to market. Is that useful?'

'Yeah, I think it's useful. It also has downsides because my fear is that people will come up with inferior products just to be the first. That will harm the reputation of the technology. Some companies seem to be willing to sacrifice quality for commercial success. That, I'm worried about.'

It's easy to imagine clean meat will be in safe hands if there are people like Mark driving it. (He prefers 'cultured' meat to 'clean' or 'cell-based' meat.) Like my first conversation with Bruce, it feels like there is no criticism of the industry I can bring up that he can't intelligently and eloquently address.

When I ask if all of this is a bubble, he says if it is, it doesn't matter. 'I'm older than most people in this field. I have a slightly more grey perspective,' he says. 'This might be one of those technologies that goes through a hype cycle, where there will be a trough of disappointment and private investors start to back out. That's the time to start a big campaign for public investment.' Mark would much prefer to be working with public money. 'This is going to be a scientific programme for the next thirty years. Even if there's a product on the market three years from now, it still requires a lot of research and tweaking. You need scientific breadth to do that, and that comes from public funding.'

When I ask if his work will encourage overconsumption, he waves the suggestion away. 'Every ageing person has increasing problems with digesting meat. It's just not physiologically possible to eat more meat than feels comfortable; there's an upper limit to it. Meat consumption in heavily industrialized countries is actually going down.'

But when I put Matthew Cole's point to him, about perpetuating the taste for meat when it might be cultural rather than natural, Mark says something I wasn't expecting.

'Meat *is* a cultural thing. Part of the appeal of meat – I'm now saying something extremely controversial, but I think there's an element to it – part of the appeal of eating meat is that you actually *have to* kill animals for it.'

'In what way? What do you mean?'

'It's supremacy over other species. Meat has always been associated with power, with masculinity, with fire, with all those things.'

And he tells me about an ad for Remia barbecue sauce that ran recently on Dutch TV: Sylvester Stallone knocks a vegetable kebab out of a skinny actor's hands, before firing a bazooka from a helicopter. 'If you want to fight like a tiger, don't eat like a rabbit,' Sly shouts in the actor's face. Then he smears a huge steak with sauce and slams it down on a table in front of him. 'You want to act like a man? Eat like a man,' he growls.

'If you're going to make meat in a lab or in a factory, with no risk involved, with no killing involved, it becomes a very wimpy version of meat,' Mark goes on. 'It becomes much more like broccoli than like a hamburger. Being a transitional product might actually help moving towards a plant-based diet.'

And I suddenly understand why meat matters so much, why it's so hard for us to let it go: meat *is* an intrinsic part of what makes men men and what makes us human, agents that dominate the world around us, top carnivores that have unequivocal power over and control of the environment.

'This is all bound up with what it means to be human, isn't it?' I say.

'Right.'

'Being human means dominating the world. And we've dominated it so well that we are destroying it now.'

'Right.'

Clean meat is going to change what it means to be human:

human beings will no longer live at the expense of animals. But if meat is cultural rather than natural, it's within our power to change our culture without relying on technology. Our culture has already changed: masculinity is no longer defined in terms of the ability to make fire and kill. Yes, clean meat could be the transitional product that weans us off killing animals in the same way that sex robots might be methadone for sex offenders. But it could also prolong our addiction, and leave us dependent on faceless multinationals for basic food. Instead of relinquishing our power to dominate animals by giving up meat, we are giving remote corporations more power to dominate us.

'Couldn't this be driving us towards a world where we are relying on very specialized technology and corporations to produce our food where we were once self-sufficient? If you are a farmer in Vietnam, you can raise your own pigs for food. In a future where killing animals is forbidden but eating them is still normal, we'll be disempowering ourselves by depending on technology.'

'Yes. And I fully agree,' he replies immediately. 'I talk about microbreweries or "microcarneries" to illustrate that you don't necessarily need to associate this technology with multinationals doing this in low-wage countries somewhere far away.'

'But it's not going to happen like that, is it.'

'It . . . You know . . . we *do* have microbreweries.'

'But people drink Heineken and Budweiser. We have microbreweries, but they are for something like 0.5 per cent of the global market.'

'Yes, but they still are there. They are now 0.5 per cent of the market, but we don't know what they are going to evolve into. But I completely agree with you, the fact of the matter is people would rather pay £4.99 than £5 for a kilo of beef, and if you want to go to £4.99 instead of £5, you need to scale up to a very large scale. And then you need to accept that it comes from very, very far away. This is consumer driven, I guess.'

'Don't you think that's a dark, troubling idea?'

'It is, but that's to accept the dark side of the human species. I'm not a big believer that we are victim to large multinationals. We *give* them the power to be those large multinationals. I tend to be very liberal about these things – if this is what's going to happen, it's presumably the will of the people. I would prefer to see microbreweries, but that's not in my hands. If Unilever wants to start culturing sausages, I cannot stop them.'

Clean meat is one of many possible futures of food, so long as we continue to eat meat. We will always have the power to not want it anymore, or to want it much less. That is where the real power lies: in harnessing our desires, rather than in mastering technology. Until we do, we will be even further removed from where our food comes from, and will feel even less responsible for it. We will be perpetuating the kind of thinking that caused the meat mess in the first place.

PART THREE

THE FUTURE OF BIRTH

Ectogenesis

CHAPTER NINE

The Business of Baby-Carrying

The Pacific Fertility Centre on Los Angeles' Wilshire Boulevard is the place where the people who have it all make their babies. The waiting room's walls are upholstered in studded cream leather, the sofas are crushed velvet in shades of mink and ivory, bowls of white orchids rest underneath crystal chandeliers. You'd be forgiven for thinking this was the changing room of a high-end bridal shop, but the images on the flat screen on one of the walls give it away: digital photos of newborns in scratch mittens, thank you notes, posed family Christmas cards, tiny heads cradled in grateful hands. The pictures of the babies scroll upwards and disappear, like bubbles in champagne.

A tall but tiny woman is sitting on my left, dressed in navy leggings and running shoes. She can't be older than twenty-five. Her cropped sweatshirt reveals tanned skin, an impossibly flat stomach and slim waist; her short, bleached hair, dark lashes and delicate jawline could only belong to a model. Her swan neck is bent over her iPhone, long fingers scrolling through Instagram, long fingernails occasionally tapping on something. On my right, another woman is waiting: ever so slightly older, but just as striking. She wears a straw-coloured beanie and no make up, and her hands are so tiny she needs both of them to hold up her jewel-encrusted iPhone case.

Dr Vicken Sahakian is finally ready to see me. I go down a

corridor lined with photo collages in black frames. There's a newborn in a Santa hat, tucked into a red Christmas stocking. There's two men with tears in their eyes, each cradling a swaddled twin in his arms.

In the twenty-five years Sahakian has been a fertility specialist he has made families for thousands of the world's most privileged people. His clients are straight, gay, young and old, and they come to him from across the globe, especially China, the UK and other parts of Europe where surrogacy is either illegal or very tightly regulated. In California surrogates are allowed to make a profit from carrying other people's children, and the legal system here is renowned for upholding the rights of intended parents over any third parties who might be involved in creating their babies. It's given the state a reputation as one of the most surrogacy-friendly places in the world.

As diverse as they are, Sahakian's clients have one thing in common: they can afford him. If you are open to using other people's eggs, sperm or uteruses and are prepared to pay, he can make anything possible.

'Money talks. If you have money, you're going to have a baby,' he tells me, less than five minutes after I have sat down opposite him at his enormous black desk in his monochrome corner office. Next to his keyboard there is a coaster with the words *Babies Are Such a Nice Way to Start People*, a plastic uterus and fallopian tubes, and a glass cube paperweight containing a laser-engraved baby.

'It's sad, but that's the case.' He checks himself. 'It isn't sad, actually – it's pretty happy. When I was in training I was almost going to forfeit this field because it was too sad. We would call nine out of ten patients and tell them, "You're not pregnant." Now it's a 180-degree change, from a technology that was marginally successful when I started out to a technology that is now almost always successful. I believe in this type of science. I believe in family balancing, gender selection, selecting out abnormal embryos, using egg donors, sperm donors. This is what I do. I *love* what I do. The ultimate goal here is bringing happiness for someone.'

As the range of fertility options open to his clients has diversified, so have their requests. A growing number of women are coming to Sahakian for 'social surrogacy': they want to have babies who are genetically their own, but they don't want to be pregnant and give birth to them. There is no medical reason for them not to carry their own babies, they would just prefer to use a surrogate. They conceive their children using IVF and then hire another woman to do the gestation and delivery part. It is the ultimate in outsourced labour.

'I don't have any issues with it,' he tells me plainly, sitting back in his grey surgical scrubs embroidered with his name, his hair slicked back and greying at the temples. 'You're a twenty-eight-year-old model or actress, you get pregnant, you're going to lose your job – you will. If you want to use a surrogate, I'll help you.' It costs $150,000 to have this sort of help, and more women than ever are prepared to pay for it. 'Five years ago I would have seen a handful a year. Now, probably twenty a year. And if I'm seeing that, there are *so* many reproductive endocrinologists in the area who are very competent fertility specialists, I'm sure they are seeing the same.'

'Do you think more women would be doing social surrogacy if they could afford it?' I ask.

'Absolutely. There's an advantage of being pregnant, the bonding, I understand that, but as a man I can't understand what it is. From experience, I can say that most women love to be pregnant. But a lot of women don't want to be pregnant and lose a year of their careers.'

Sahakian doesn't have a typical client. 'I work with everybody.' But there are Hollywood stars, household names he is too discreet to tell me: 'You won't hear it from me but of course you would have heard of them.' The women who come to him for social surrogacy aren't the big stars, he says; if you have real heft in Hollywood you have the leverage to call the shots when it comes to schedules, and you can be confident that your job will still be waiting for you if you have a career break to have a baby. The typical candidates are ascending in the entertainment industry but haven't yet made their names.

'They tell me point blank, "If I get pregnant, I will lose my part." "I work, I don't have time because of work." "I model, I act, I look good like this and I don't want to disfigure my body."'

I wince. 'Do you disfigure your body when you get pregnant?'

'I've never been pregnant,' he shoots back with a sparkling grin, and I might be imagining it, but I swear he takes a quick glance at my torso, as if to gauge whether a former pregnant person has asked the question. 'You are definitely disfiguring your body *while* you're pregnant, for that duration, and then if you don't do the necessary exercises it's going to take you a while to get back to normal. There's definitely some truth about pregnancy changing your body. Your pelvic bone opens up, you accumulate fat, you accumulate discoloration that doesn't go away. Things change. I'm not saying that's a reason to use a surrogate, but it is for some people.'

He shifts in his big leather swivel chair and tries another approach. 'I make the analogy of plastic surgery. If you criticize somebody who's had a breast augmentation then you are certainly going to criticize somebody who wants to do social surrogacy. One is saying, "I don't feel comfortable with my body, it's psychologically an issue for me and I want to fix it." The other is saying, "I don't want to disfigure myself."'

Not all of his social surrogacy clients are models and actresses; some just have demanding careers that would be very inconvenienced by pregnancy. 'I have many clients who say, "I can't, I have to travel, I don't want to wait any longer, I am getting older and my career in the next two or three years is critical, I travel all the time." It's an honest argument.'

'Do women generally do it for aesthetic or professional reasons?'

'Professional, I would say. "I don't have time because of work" is common, followed by physical appearance.'

Men get to be parents without it disrupting their lives very much, no matter how high profile or demanding their jobs. They often don't even need to consider the impact having a baby will have on their careers, even at the most critical times: former Lib Dem leader Charles Kennedy's son, Donald, was born during the 2005 general election campaign. Mo Farrah's wife, Tania Nell, gave birth to twins three weeks after he won two gold medals at the 2012 Olympics.

'What do the partners of the women who come in for social surrogacy think?'

It's clearly the first time Sahakian has ever considered this question. 'You know, I never bring that up! I never ask that question.'

'But do they come in with their partners?'

'Yeah, yeah, of course.'

Sahakian tells me that his years of working in the fertility field have turned him into a feminist. 'I am *such* a feminist, because every day I see how prejudiced this society is, how male chauvinistic it is. You women are judged. I am very proactive when it comes to women and I believe there is a double standard.'

'Do you mean that men get to have babies and keep their careers while women often can't?'

'Oh, more than that. If you are a sixty-two-year-old man and you come here with a thirty-eight-year-old woman, no one asks you why you're having a kid at sixty-two. If you come here as a fifty-five-year-old woman trying to have a kid, they'd tell you you're old, you're a grandma, you're crazy. Larry King was, what, seventy-five when he had kids?' King was actually sixty-five, but Sahakian's got a point. He himself is fifty-six, with a wife who is twenty years younger and two children under six, who look down on us with perfect smiles from the frames on his wall.

The American Society of Reproductive Medicine has guidelines that say that gestational carriers – surrogates who carry babies conceived through IVF, with eggs that aren't their own – should only be used when there is a medical need. But Sahakian doesn't have concerns about going against those guidelines.

'You can define medical reasons broadly,' he says, casually. 'And also I understand that it's controversial – you wouldn't be here otherwise. It's borderline unethical for some people, but so what? So what. Put yourself in the shoes of a twenty-six-year-old model who is making her living by modelling swimsuits. Tell me something – is it that unethical to say, "Let's not destroy this woman's career"?'

'Couldn't she wait until she's older to have a baby?'

'Yes. But what if you want to have a child now, and you don't want to be maybe forty when you have a kid? I don't think I'm doing anything unethical by helping those couples. In this field, in Los Angeles, you can't judge those clients. This is the Wild West. Twenty years ago helping a gay couple was taboo – it still is, in Arkansas. We are so in the infancy of all this here.'

'You don't have any ethical qualms at all about doing this?'

'You're talking to the wrong person,' he chuckles. 'I walk the edge, you know.'

And yes, I do know. Sahakian has a reputation for pushing boundaries that he clearly relishes; it's given him a degree of notoriety that drives his business. In 2001, he helped Jeanine Salomone get pregnant with a donor egg and give birth at sixty-two. She is the oldest French woman on record to have a baby. A scandal erupted in France, where giving fertility treatment to post-menopausal women is illegal, after it emerged that the biological father of the baby she gave birth to was actually her own brother, Robert. He may have had limited capacity to consent to his sperm being used to conceive a child; he was living with a brain injury he sustained after shooting himself in the chin in a failed suicide attempt a few years previously. French journalists suggested that their son, Benoit-David, might have been conceived to secure an inheritance from Jeanine and Robert's wealthy mother. The press descended on Sahakian, who said the siblings had presented themselves in his consulting room as a married couple, and Jeanine had lied about her age.

I knew all this before I arrived in Los Angeles, but Sahakian brings it up before I can. In fact, he brings it up when I ask him why clients come to him.

'I got the oldest woman on record in France to carry a baby and have a baby at sixty-two. You can Google that story and find out the details. Basically, there was a social stigma surrounding that story.'

'They were brother and sister.'

He nods. 'They were brother and sister. So I was put on the map – the message from that was, "Hey, this guy can get a sixty-two-year-old woman pregnant." So I had everybody over fifty calling me in the 2000s.'

Then, in 2006, Sahakian became responsible for the oldest woman *in the world* to give birth. Maria del Carmen Bousada, a retired sales assistant from Spain, had her twin boys, Christian and Pau, the week before her sixty-seventh birthday. Bousada was diagnosed with cancer less than a year later, and died in 2009, leaving her sons orphaned at only two and a half.

'That woman from Barcelona is in the Guinness Book as the oldest woman on record to give birth, actually,' he says with a pride that feels grotesque.

'Do you like having a reputation for pushing boundaries?' I ask.

'I didn't push the boundaries with the Spanish woman. She lied about her age, she said she was fifty-seven. She was sixty-seven. She forged documents, she forged her medical records. With the French people, they came in as husband and wife with the same last name – we had their passports. We don't ask for a marriage certificate, we don't ask for birth certificates. Which doctor asks for a birth certificate?'

'That sixty-seven-year-old woman died and left her kids without a mother,' I say. 'What about them?'

'That's why I wouldn't treat a sixty-seven-year-old woman,' he replies, without missing a beat. 'She was a perfectly healthy fifty-seven-year-old. She died from cancer so she didn't have a pre-existing condition. You can get cancer at twenty-eight.' He's now cut his upper age limit down to fifty-five, but he still doesn't ask his clients for conclusive proof of age.

Sahakian says none of his social surrogacy clients will speak to me. 'They have nothing to gain.' They're doing this to save their careers and have no interest in becoming the poster girls for this new way of having it all. It's taboo to say you want a baby but are not prepared to carry it, so much so that he's had a few clients who actually pretended to be pregnant, reassured in the knowledge that their pre-'pregnancy' bodies would be there for them as soon as the baby arrived. 'You can buy artificial, prosthetic bellies, you know. You can buy them in different sizes. There's a reason why that's there.'

*

Some women want children but don't want to be pregnant. It's a rarely spoken but undeniable fact. It's considered unnatural – heretical, even – to want babies without pregnancy, but that doesn't stop some women from thinking about it, and even expressing it, under the veil of anonymity. An 'Am I Being Unreasonable' thread on the parenting site Mumsnet, entitled 'If you had money to burn, would you use a surrogate?', asked users if they would 'pay for an American surrogate if you simply didn't want to wait/go through the pregnancy'. At least seven women said they would. 'Oh god yes. I had horrible HG [hyperemesis gravidum] with both my pregnancies but even putting that aside it's not an experience I savoured,' said one. 'Yes I would. Pregnancy is horrible!' said another. 'In a heart-beat,' said a third.

But most of the responses in the thread were negative and outraged. There's a tacit acceptance that a woman who wants to raise a child but doesn't want to give birth to it isn't fit to be a mother, because if she isn't willing to undergo the initial sacrifice of giving up her body to a baby, she won't ever be able to put the child first. This makes superficial sense, until you remember that fathers find a way to put their children first without giving up their bodies to them; they have to do this, by default. The physical sacrifice of bearing a child doesn't necessarily make you an attentive parent, and to assume it does is to claim that men can never be as devoted to their children as mothers are. And there are plenty of mothers who are more than happy to go through pregnancy and birth, but are unwilling to put the baby first when it arrives.

There are serious reasons why women might not want to go through pregnancy. As much as some of Sahakian's clients might buy prosthetic bumps and pretend to be pregnant while using a surrogate, a much larger number of women do the opposite: they conceal their pregnancies for as long as they possibly can, in the knowledge that being pregnant is going to cost them dearly. Despite widespread legislation to prevent it, pregnancy discrimination is a reality for women today across the globe. A study by the UK's Equality and Human Rights Commission found that one in five British mothers have experienced harassment or negative comments after revealing their pregnancy at work, and 54,000 women a year are pushed out of their jobs due to pregnancy or maternity leave. In the US, the National Partnership for Women and Families says nearly 31,000 pregnancy discrimination charges were filed with the Equal Employment Opportunity Commission between 2010 and 2015. Women in all industries, in every US state and from every ethnicity, experienced pregnancy discrimination in the workplace.

Only a tiny minority of women worldwide can afford to hire a surrogate, but many more might have valid reasons for thinking twice about carrying their own children. Some of America's biggest tech and media companies are already paying for their female members of staff to freeze their eggs so they don't have to worry about their fertility clock ticking away while they sit at their desks. Is there a future where companies will support mothers looking for someone else to carry their babies so pregnancy doesn't interrupt their work?

Look closely at the wording on any number of California-based fertility clinics' websites and you will see that surrogacy for non-medical reasons is on the cards. 'Couples and individuals who are unable to have a baby on their own either biologically *or through intention* can still build and grow a family thanks to surrogacy,' says the website of Growing Generations (with my emphasis). 'From medical to emotional to logistical and more, the indications for gestational surrogacy can vary significantly,' says the Los Angeles Reproductive Center's surrogacy page.

I ring up at least ten different California-based fertility clinics, asking if they had clients who'd be prepared to talk to me about why they chose social surrogacy. They all repeated some version of Sahakian's line: this is not about vanity, it's about the pressure women are under to maintain their careers at the same time as becoming parents, and women know it's not acceptable to admit you've used a surrogate for non-medical reasons, so no one will talk.

A more nuanced picture begins to emerge when I speak to people in the industry outside Hollywood. An assistant at a clinic in San Diego tells me their social surrogacy clients tend to be single women in high-powered corporate careers who would risk losing their jobs if they had debilitating morning sickness or were put on bed rest; carrying a baby themselves would not only risk their bodies and health but the livelihood and income their child will eventually depend on. A fertility doctor tells me that 80 per cent of her social surrogacy clients are Chinese, because of a 'cultural thing' in China where a uterus is considered to be old after one pregnancy. A fertility psychologist who used to run her own surrogacy agency says she worked with a woman who was campaigning for political office and desperately wanted a child; she had to be out on the campaign trail or risk jeopardizing everything she had ever worked for, so she hired someone else to carry her baby for her.

But what about the surrogates, the uterus-bearers expected to 'disfigure' themselves so someone else doesn't have to? How do they feel about potentially risking their lives to give a baby over to someone who has no medical reason not to carry it herself? Well, generally they have no idea that's what they are doing. Lori Arnold, a fertility specialist from San Diego who runs both a clinic and her own surrogacy agency to provide carriers for her clients, tells me that 'the surrogates really don't know the medical point of why the intended parents are seeking surrogacy. If they asked, if we had permission from the intended parent, we would tell them. But it's a personal medical decision that I do keep private and confidential.'

*

Surrogacy is never the easy option. Even with the most willing surrogate, the most professional fertility doctor and the most fastidious paperwork, surrogacy is the most physically, emotionally and legally messy form of third-party reproduction. But it is the only solution to the problem of baby carrying that humans have ever had.

Traditional surrogacy – where the surrogate is the genetic mother of the baby she carries but gives up her parental rights – has been around from the Book of Genesis to *The Handmaid's Tale*. Genesis 16 tells the story of how Sarah and Abraham were having problems conceiving an heir. Sarah told Abraham to go to her Egyptian slave, Hagar, 'that I shall obtain children by her'. It didn't end well: as soon as Hagar discovered she was pregnant with a son, Ishmael, 'she looked with contempt on her mistress', and when Sarah had her own biological son, Isaac, fourteen years later, she cast Hagar and Ishmael out into the desert.

Although traditional surrogacy has existed for millennia in one form or another, it was usually shrouded in secrecy because of the taboo of infertility, the stigma of illegitimacy and the simple mechanics of what's involved in making a baby this way. Artificial insemination took some of the ickiness out of traditional surrogacy, but it has its own dark history: the first recorded case took place in Philadelphia in 1884, when Professor William Pancoast helped an infertile man and his wife conceive. Pancoast used a rubber syringe to inject fresh sperm from one of his 'best-looking' students into the cervix of the woman after she had been knocked out with chloroform. She gave birth nine months later. She was never told how she had conceived, or that her husband was not her baby's biological father.

The technique Pancoast pioneered changed what it means to make babies: getting pregnant no longer had to depend on heterosexual sex. This has been great for lesbian and gay couples, although, of course, gay men still need women to carry their babies for them. Traditional surrogacy (as opposed to gestational) still remains an option today; it's the cheapest way of having a child with a surrogate, and if the surrogate is related to one of the intended parents it allows them to have a further genetic link with their babies.

When Louise Brown, the first baby conceived through IVF, was born in Oldham in 1978, a new era of baby-carrying possibilities was born along with her. Not only was conception no longer dependent on sex, it could happen outside the womb, meaning that it became possible for a woman to be pregnant with another woman's child. The first baby conceived through egg donation was born in 1982, and in 1985 the first successful case of gestational surrogacy was recorded. There could now be a distinction between a genetic mother and a birth mother. For the first time, motherhood became fragmented.

Since the 1980s we have become gradually more willing to accept that a birth mother can be a different person from a genetic mother. It's difficult to quantify the rise in gestational surrogacy with any accuracy, but in 2014 the *New York Times* estimated that three times as many babies were born through gestational surrogacy in the US compared to a decade earlier, and in 2018 it was estimated that in Canada, where only altruistic surrogacy is legal, it had increased by 400 per cent since 2008. The rise of gay marriage has led to a greater acceptance of gay parenting, at a time when fewer babies are being put up for adoption at birth. Single men have begun looking to surrogates in the same way that single women might consider using sperm banks to have children on their own. Surrogacy is increasingly the way that modern families are made for people who can't or won't bear their own children, and gestational surrogacy has become far more popular than traditional surrogacy: it is a safer bet, as the embryos have already been created by the time they reach the surrogate's womb, and many in the fertility field say it is legally and emotionally easier than asking a woman who has just given birth to her own genetic offspring to immediately hand it over.

But all forms of surrogacy come with serious legal and ethical challenges, whether they are traditional, gestational, commercial or altruistic. You might imagine the main issue would be surrogates getting attached to the babies they are carrying and refusing to give up them up, but it's actually far more likely that the intended parents will change their minds and decide they no longer want an already-gestating child. Surrogates have been asked to terminate pregnancies against their will when intended parents split up, or when anomalies and disabilities are detected in the foetus; they have even been asked to abort 'surplus' babies when too many embryos implant successfully. There are far too many documented cases of this.

In 2014, an international scandal erupted when a gestational surrogate from Thailand, Pattaramon Janbua, was trying to raise money to help her bring up a baby she said had been abandoned by his Australian intended parents because he had been born with Down's syndrome. Pattaramon had been pregnant with boy–girl twins, and a scan at seven months revealed that the boy, Gammy, had birth defects. Her clients, David Farnell and Wendy Li, asked her to abort him. Pattaramon refused, and said the Farnells came to Thailand after the birth to take their daughter, Pipah, but not Gammy. It later emerged that David Farnell was a convicted paedophile who had served time in jail for assaulting two girls under the age of ten. In 2016, courts in Western Australia ruled that the Farnells had not abandoned Gammy; they had wanted to keep both babies, but Pattaramon did not want to give Gammy to them. Pipah is not allowed to be alone with her father but is staying with the Farnells, and Gammy is remaining with Pattaramon. The judge said the case 'should also draw attention to the fact that surrogate mothers are not baby-growing machines, or "gestational carriers" . . . They are flesh and blood women.' Thailand banned commercial surrogacy for foreign parents in 2015.

International commercial surrogacy is fraught with its own special blend of ethical problems. Like any kind of outsourced labour, it's the poorest and least enfranchised people who bear the brunt of the market. Fertility tourism from the UK used to be a growing industry in India, where poor and often illiterate gestational surrogates were regularly made to stay under close surveillance in dorms for the nine months of their pregnancy, and intended parents were allowed to dictate what they ate and whether they were allowed to have sex. Complete packages, including the surrogate fee and all medical bills, started from as little as $10,000. When India finally outlawed international surrogacy in 2015 the industry was estimated to be worth $500 million a year. Now Ukraine is the go-to destination for cut-price gestational surrogacy, but it's not uncommon for Ukrainian surrogates to be abandoned without payment if they miscarry, or subjected to more caesarean sections than is medically safe. Multiple embryos are transferred to maximize the chance of a successful pregnancy, with little thought about how the surrogate might cope with carrying triplets or quads.

No matter how many happy surrogates around the world argue that they are carrying other people's children in order to give the gift of parenthood to the people who want it most, surrogacy by definition depends on using a woman as a vessel, an incubator, and then expecting her to give up any right she might have to the baby inside her. It depends on exploiting women's reproductive potential, whether or not they consider themselves exploited. In December 2015, the European Parliament condemned all forms of surrogacy, on the grounds that it 'undermines the human dignity' of women, and singled out gestational surrogacy specifically, because it 'involves reproductive exploitation and use of the human body'.

But banning surrogacy won't remove the demand for it. It is too late; the possibility of having a genetically related baby without pregnancy has opened up a new world to both men and women, one that cannot just be waved away. And new reasons for demanding pregnancy-free parenthood emerge every year, as Sahakian's bulging client list shows.

*

There is no doubt that the fertility specialists who offer social surrogacy are at the most extreme end of fertility treatment. But those same doctors have led the way when it comes to creating families for older women, same sex couples and single men and women. Could this be another barrier they are breaking down, blazing a trail that the rest of the world will one day follow?

Sahakian likes to think so.

'You say twenty years ago LA was the Wild West in terms of gay surrogacy,' I say. 'Do you think twenty years from now that's how people will think of social surrogacy?'

'Twenty? No, a couple of years from now. We're already almost there. Surrogacy isn't taboo anymore. In the UK, you are so far behind us. Thank God – it's good for business! But that's going to change.'

The strange thing is, the secrecy surrounding social surrogacy makes it more likely that more people will want to have their babies this way. The women who use Sahakian's clinic are the people who the rest of us are supposed to look at and aspire to be.

'Aren't you creating an illusion that it's possible for these women to have the career and the body and the family, to have it all, when it isn't?'

He shrugs it off. 'I don't think it's a social problem. I can see both sides, but I'm not going to judge. If you want a baby and have someone else carry it, you are helping two people – you are going to have the kid and the surrogate is going to make money helping you.'

I don't really buy this. Given a choice, I'm sure Sahakian's social surrogacy clients would rather not have to go through the messy and complicated business of surrogacy, but if they want a baby and they don't want to carry it themselves, they will have to put up with it.

For now.

Because the drive to improve the technology that has changed the meaning of motherhood rumbles on. First babies didn't need to be conceived through sex, then they didn't need to be created inside their mother's body. What if we could have babies without anyone being pregnant?

CHAPTER TEN

The Biobag

The lamb is sleeping. It lies on its side, eyes shut, ears folded back and twitching. It swallows, wriggles and shuffles its gangly legs. Its crooked little half-smile makes it look particularly content, as if dreaming about gambolling in a grassy field somewhere. But this lamb is too tiny to venture into the outside world. Its eyes cannot open. It is hairless; its skin gathers in pink rolls at the base of its neck. It hasn't been born yet, but here it is, at 111 days' gestation, totally separate from its mother or any other animal, alive and kicking in a research lab in Philadelphia. It is submerged in fluid, floating inside a transparent plastic bag, with its umbilical cord connected to a nexus of bright blood-filled tubes. This is a foetus growing inside an artificial womb.

Here it is, two weeks later, at 135 days' gestation, almost full term. The lamb nearly fills the entire space; its flat nose presses up against the corner of the bag. It is plumper, whiter, fluffier, covered in fine coils of wool, with a puff of a tail: definitely a lamb now, but still a foetus. In another two weeks, the Ziploc bag will be unzipped, the umbilical cord will be clamped and the lamb will finally be born.

When I first see images of the lamb in the bag on my laptop, I think of the foetus fields in *The Matrix*, where motherless babies are horrifically farmed in pods on an industrial scale: human Cowschwitz. But this is not a substitute for full gestation. California's booming surrogacy industry can rest easy, for now. The Philadelphia lambs didn't grow in the bags from conception; they were taken from their mothers' wombs by caesarean section and then almost immediately submerged in the biobag at a gestational age equivalent to around twenty-three to twenty-four weeks in humans. This isn't a replacement for pregnancy yet, but it is certainly the beginning of it. Birth may one day be as simple as opening a Ziploc bag.

The team who made these artificial wombs say they are driven only by the desire to save the most vulnerable humans on earth. Emily Partridge, Marcus Davey and Alan Flake are neonatologists, developmental physiologists and surgeons who work with extremely premature babies at the Children's Hospital of Philadelphia (CHOP). After three years of tweaking and refinement, their latest prototype – 'the biobag' – is designed to give babies born too soon a greater chance of survival than ever before.

The biobag was born into public consciousness in April 2017, when the CHOP team published their research, along with images of the lambs, in the journal *Nature Communications*. The paper describes the four different artificial womb prototypes CHOP tested on a total of twenty-three lambs before settling on the biobag design. (Sheep are the go-to animal models in obstetric research because they have a long gestation period, and their foetuses are around the same size as ours.)

'In the developed world, extreme prematurity is the leading cause of neonatal mortality and morbidity,' the paper begins. 'We show that fetal lambs that are developmentally equivalent to the extreme premature human infant can be physiologically supported in this extra-uterine device for up to 4 weeks [. . .] With appropriate nutritional support, lambs on the system demonstrate normal somatic growth, lung maturation and brain growth.' They had found a way to gestate foetuses outside maternal bodies; foetuses that would eventually become lambs no different from those which had grown in the wombs of pregnant ewes.

CHOP's communications department released a very slick short film to coincide with the paper's release. I imagine this was intended to focus the inevitable international press attention on the therapeutic benefits of the biobag instead of the freaky lamb images. Entitled *Recreating the Womb*, it looks very much like a corporate video, and there is not a foetus to be seen throughout its entire nine-minute duration. There are neat diagrams of lambs in biobag systems, and some slightly awkward staged B-roll footage of Partridge, Flake and Davey pretending to do lambless lamb research in a pristine lab, set to some twinkly piano music intended to inspire awe and wonder. There are some heartbreaking clips of superpremature babies in CHOP's neonatal intensive care unit (NICU): impossibly small humans covered in tubes, tiny fingers with cracked, flaking skin, breathing tubes taped to gasping little mouths. And then there are some glossy set-piece interviews with the research team themselves in white lab coats: carefully edited, backlit and shot in a studio. A longer version was released at the same time as the shorter promotional film, and includes extended interviews with the team.

'In the future, we envision the system will be in the NICU and it will look pretty much like a traditional incubator. It will have a lid that will be able to move up and down,' Davey says, his accent half Australian, half American; he was born in Melbourne and came to CHOP in 1999. 'Inside that warmed environment will be the baby inside the biobag. We'll have amniotic fluid next door to the incubator, which will be pumped through the biobag,' he adds, in the extended video.

The biobags would be kept in a darkened environment to

mimic the human womb, but the babies would be visible as never before. 'Parents would see a lot more than they see during a normal pregnancy. We'd have a dark-field camera in the unit so they can actually look at their foetus in real time, see their foetus move and breathe and swallow and do all the things that foetuses do,' says Flake, the most senior member of the CHOP team. 'There will also be an ultrasound unit. That's really how we'll do physical examinations on the foetus, since we can't touch the baby like you can in a preterm infant. We'll do ultrasound and look at its physiologic well-being at least once or twice a day.'

We do like to monitor our babies. In a world where parenthood increasingly begins with fertility-tracking apps followed by what-to-expect-when-you're-expecting apps and then apps that track every newborn feed and bowel movement, along with video monitors that measure your baby's vital signs and stream everything to your phone in all its night-vision glory, this will fit right in.

'It never fails to strike me what a miracle it is to see this foetus that is clearly not ready to be born enclosed in this fluid space, breathing, swallowing, swimming, dreaming, with complete detachment from the placenta and from mom. It is an awe-inspiring sight,' says Partridge, smiling with her eyes shut just like the lamb in the video, and shaking her head as if she can't believe what she's managed to do.

It's a team effort, but Partridge talks like the biobag is her baby. She is the most junior member of the team and the only woman; she came to CHOP from Toronto as a research fellow. In interviews with Canadian broadcaster CBC on the day the paper was published, she takes ownership of the whole concept. 'I pitched this idea, really, with the belief that this offered an unprecedented opportunity to improve what we can do for these babies,' she says. She describes looking after the lambs in the bags like a mother beside her newborn's cot: 'rolling out a sleeping bag and camping out beside these lambs for weeks and weeks at a time.'

In the CHOP video, Partridge talks us through the two key components of the biobag that act as substitutes for a mother's body. The placenta is replaced by a circulatory system, 'a device in which blood flows and carbon dioxide is removed and oxygen is added to that blood.' This is an oxygenator plugged into veins in the lamb's umbilical cord, which also delivers nutrients and any necessary medication the lamb might need. (It actually performs exactly the same function as the woman in the lab coat at JUST pipetting the medium in and out of the seed trays to allow the chicken cells to proliferate.) There are no mechanical pumps to drive the blood through it, because even gentle artificial pressure could overload the lamb's tiny heart. The flow of blood is instead pumped entirely by the beating of the foetus's heart, just as it would be in the womb.

'The other component is recreating the womb itself, and that's really the fluid environment, which we have recreated with a soft, bag-like structure,' Partridge continues. 'It's meant in some ways to swaddle and keep the foetus supported physically the way that it would be in the uterus.' The plastic bag acts like an amniotic sac filled with warm, sterile, lab-made amniotic fluid that the lamb breathes and swallows, just like a human foetus would. This fluid flows in and out of the biobag through tubes in two small, watertight apertures. The team went through 300 gallons of the stuff a day during their experiments.

The biobag is needed because of the fallibility of the uterus. Normal pregnancy is forty weeks; any baby born before thirty-seven weeks is considered premature. At twenty-three weeks – a little over five months – the mother has only just passed the halfway point of her pregnancy. One per cent of all babies born each year in the US are born as premature as this, Flake says. That twenty-three- to twenty-four-week figure is totemic: it's the border of viability, the age after which modern medicine currently has a hope of keeping babies alive when they are born early, and the point at which doctors will attempt to resuscitate a newborn. The NHS currently uses twenty-four weeks as the border of viability, and a baby born dead at twenty-four weeks is classed as a stillbirth, whereas a dead baby born at twenty-three weeks and six days is still called a miscarriage. It is a brutal boundary.

In countries with good hospitals there is currently a 24 per cent chance of keeping a baby born at twenty-three weeks alive. But 87 per cent of those that make it will go on to experience major complications that dominate their lives, like chronic lung disease, bowel problems, brain damage, blindness, deafness and cerebral palsy. More extremely premature babies are surviving in wealthier countries: between 1995 and 2006 in England, there was a 44 per cent increase in babies born before twenty-four weeks living long enough to receive neonatal care. But we aren't getting better at avoiding the problems associated with premature birth at this stage, and the number of children growing up with chronic conditions associated with prematurity has also increased dramatically. Preterm birth is the greatest cause of death and disability among children under five in the developed world.

Incubators deal with some of the functions a premature newborn needs help with, but they don't allow for the process of gestation to continue. They provide warmth and humidity, but not nutrients – that's why the babies inside them are covered in catheters and cannulas that deliver what they need to survive and grow, and also why they have to be sedated: to stop them trying to pull the tubes out of themselves. Ventilators keep premature newborns alive by breathing on behalf of their underdeveloped lungs, but also increase the chance of infection, stop the lungs from developing properly and can potentially damage whatever delicate lung tissue is already there. Instead of supporting the newborn as it tries to survive outside its mother's body, the biobag treats the baby as a foetus who has not yet been born.

'If it's as successful as we think it can be,' Flake says in the promotional video, 'ultimately the majority of pregnancies that are predicted at risk for extreme prematurity would be delivered early onto our system, rather than being delivered premature onto a ventilator.'

I have to listen to this comment a few times. Is he saying that women *at risk* of going into very early labour would have a caesarean *just in case*, so their babies could be transferred into the artificial womb for the rest of their gestation?

But then he continues: 'With that, we would have normal physiologic development, and avoid essentially all of the major risks of prematurity, and that would translate into a huge impact on paediatric health.' The shots in the video are now of chubby babies sitting up and giggling, a gap-toothed six-year-old grinning, a young woman breaking into a slow-motion smile. If a biobag could mean a healthy future for so many babies instead of one of illness and disability, who could deny it to them?

This is how CHOP approaches all of the potentially massive controversies their artificial womb brings up: by focusing on paediatric health, on gurgling babies, to the exclusion of everything else. There are no ewes in the film or the scientific paper, and no input from mothers. The researchers want their device to be seen as ethically unremarkable, to be about helping sick babies and nothing more. 'Our goal is not to extend the current limits of viability, but rather to offer the potential for improved outcomes for those infants who are already being routinely resuscitated and cared for in neonatal intensive care units,' the paper says, carefully. Extending the current limits of viability would create an ethical minefield. The legal abortion limit in the UK was brought down from twenty-eight to twenty-four weeks in 1990 because advances in neonatal care meant that foetuses born between twenty-four and twenty-eight weeks were more likely to survive. If artificial wombs are helping ever-smaller babies survive, that can have enormous implications for women. But women are not mentioned in CHOP's work.

The closing sentences of the scientific paper show how absent women are from their considerations: 'Our system offers an intriguing experimental model for addressing fundamental questions regarding the role of the mother and placenta in fetal development. Long-term physiologic maintenance of a fetus amputated from the maternal–placental axis has now been achieved, making it possible to study the relative contribution of this organ to fetal maturation.'

As much as CHOP's communications department want to emphasize that the biobag is a therapeutic tool for very sick, very tiny babies, the people who made it are keen for the scientific community to know that they have managed to 'amputate' foetuses from the mother and placenta, meaning that the 'relative contribution' of pregnant mothers and their organs to how babies grow can be studied. And perhaps, eventually, be deemed entirely replaceable by technology.

As I get to the final minutes of the promotional film, it's beginning to look more like the JUST chicken video, a familiar story of good old American grit, tenacity, resourcefulness and entrepreneurship that might just save the world. Davey and

Partridge describe how the prototypes for what became the biobag developed. 'For the first few generations we used a lot of supplies from plumbing, piping, beer stores,' Partridge grimaces. 'We did not have grants at the time, so it really required a bit of innovation to create the first prototype really from nothing.'

'Thomas Edison said to be an inventor all you need is an imagination and a pile of junk. Essentially, that is the story of this system,' says Davey. 'Sometimes we ended up going to Home Depot, we ended up going to Lowes and to Michaels, we would bring these objects back to the lab and glue them and melt them all together.'

By the end of the CHOP video, Partridge is beaming with pride. 'This certainly is a project that would have sounded more like science fiction than a reality, but over three years of really doggedly pursuing it, and refusing to accept setbacks and limitations, it has become a very real therapeutic tool.'

But this is not just a therapeutic tool: it's an invention that will one day go on the market, a commodity, and CHOP wants to protect its intellectual property. After a Google deep dive I find an application to patent the biobag, filed in 2014, long before the scientific paper was submitted, and it's probably the most revealing thing the team has ever put into the public domain. There's no coyness about extending the limits of human viability in this document: it explicitly says that the possible 'subjects' who will use the invention include 'pre-viable foetuses (e.g. 20–24 weeks)'.

There are some touching details in the patent application that didn't make it into the scientific paper or the promotional video. Partridge, Flake and Davey gave their little lambs names, in their early experiments, at least. There was June, Charlotte, Lily, Little Alan, Eddie, Willow, Seinne, Bowie, Iggy and Manson.

Most of them were killed at birth so CHOP could study their organs, but a few lucky ones were allowed to live, and were bottle-fed by the team. Iggy did particularly well and was 'successfully delivered from the artificial placenta and transitioned to postnatal life [. . .] The animal displayed appropriate growth and development over eight months before transport to a long-term adoptive facility.' The final photo in the patent application is of a sprightly lamb inside a shed, looking over one shoulder to face the camera, as if striking a pose.

It is probably because of this patent that I have to rely on CHOP's online promotional videos and research paper to describe the biobag and all the work that went into it. I am not allowed to go to Philadelphia to see the team's work with my own eyes. I so nearly was: Alan Flake told me I was welcome to come; we fixed a date and time for my visit. I was about to book my flight when I thought I should really let CHOP's press office know I was coming; you can't just stroll into a children's hospital without the right permission. I had a very friendly forty-minute conversation with the CHOP press officer, who seemed keen to have me too, but told me to hold off booking anything until they got the green light from CHOP's legal department, which would take a couple of days. It sounded like a formality.

But days became weeks, and the flights were getting more expensive, and for some reason the friendly press officer now wouldn't answer my calls or return my messages. Then, finally, a very brief email dropped into my inbox. 'I'm very sorry to report that CHOP is going to pass on this opportunity,' the press officer wrote. 'I really enjoyed talking with you, and was hopeful we could make this work but could not. I apologize for the confusion and delay getting you a final answer. Thank you for your interest in this research.'

It took several more emails to get a hint of why I was

suddenly unwelcome. Flake apologized and said it was out of his hands. They want to be able to put babies inside the biobag within a couple of years, and the prospect of my visit made the legal department twitchy. 'There is a lot of caution to do anything that could jeopardize FDA approval,' the press officer finally told me. CHOP didn't want to threaten the medical and commercial future of their invention by talking to a journalist too soon. Their focus, at the moment, is getting their artificial womb to market.

When the biobag does hit the market, it will be only the latest and most literal manifestation of how pregnancy is becoming externalized. In any pregnancy in the developed world a woman is routinely prodded and probed, scanned vaginally and abdominally, has her blood taken and tested so the form, growth and DNA of her baby can be analysed. If something is suspected to be wrong with the foetus she will be probed some more: large needles will be inserted through the skin and muscle of her abdomen and into her uterus to sample cells from her placenta or her amniotic fluid, which can be subjected to further DNA testing. Even if everything is going well, it's taken for granted that a pregnant woman will be strapped to foetal heart monitors and blood pressure gauges and have the size of her cervix routinely measured while she's in labour. In an age when being a good parent means being as attentive as possible even before birth, we want better access to the babies growing inside pregnant women, better ways of measuring them and putting them under surveillance, so we can do the best for them even before they enter the world. Women's bodies are almost getting in the way.

*

'Ectogenesis' – reproduction outside the human body – was coined by British scientist J. B. S. Haldane in a lecture he delivered to the Cambridge University Heretics Society in 1923. Haldane imagined an essay written by a Cambridge University student of the future describing the great biological inventions engineered since Haldane's time. 'We can take an ovary from a woman, and keep it growing in a suitable fluid for as long as twenty years, producing a fresh ovum each month, of which 90 per cent can be fertilized, and the embryos grown successfully for nine months, and then brought out into the air,' his imaginary future essayist wrote. 'France was the first country to adopt ectogenesis officially, and by 1968 was producing 60,000 children annually by this method.'

Haldane was interested in ectogenesis for its social engineering potential, at a time of slowing birth rates; in 1923, eugenics was not yet considered a despicable idea. 'Had it not been for ectogenesis there can be little doubt that civilization would have collapsed within a measurable time owing to the greater fertility of the less desirable members of the population in almost all countries,' he said. Haldane concluded that the complete separation of reproduction from sex would mean 'mankind will be free in an altogether new sense'.

Churchill's 'Fifty Years Hence' article from 1931 had as much to say about ectogenesis as lab-grown meat. 'There seems little doubt that it will be possible to carry out in artificial surroundings the entire cycle which now leads to the birth of a child,' he wrote of his imagined 1981.

Churchill was writing only a year before Aldous Huxley published *Brave New World*. Huxley took a lot from his friend Haldane's ideas, but turned them on their head: his brave new world of 2540 was a dystopian nightmare, where reproductive technology was a form of social control. Human beings were mass-produced in bottles lined with pigs' peritonea, which spent 267 days on a production line conveyor belt in the Central

Hatchery. 'One by one the eggs were transferred from their test tubes to the larger containers; deftly the peritoneal lining was slit, the morula dropped into place, the saline solution poured in,' Huxley wrote. 'The procession marched slowly on; on through an opening in the wall, slowly on into the Social Pre-destination Room.' Here, embryos were turned into humans of different social classes: some were starved of oxygen to give them brain damage, so they were content with menial work; others were kept in freezing conditions to give them an aversion to cold, so they were happy to become miners in the tropics. Huxley's view of ectogenesis has come to dominate: its place in our collective imaginations has been a dark trope of science fiction ever since.

In the real world, the possibility of having a baby without a womb began to represent a new frontier of freedom. In the 1970 feminist classic *The Dialectic of Sex*, Canadian radical feminist Shulamith Firestone argued that the biological division of labour in natural reproduction forms the basis of male domination over women. Her 'first demand for any alternative system' was 'the freeing of women from the tyranny of their biology by any means available, and the diffusion of the childbearing and childrearing role to the society as a whole, to men as well as women.'

The manifesto of the UK's Gay Liberation Front, first published in 1971, said ectogenesis had potential to emancipate both men and women by erasing the natural distinctions between them. 'We have now reached a stage at which the human body itself, and even the reproduction of the species, is being "unnaturally" interfered with (i.e. improved) by technology,' it reads. 'Today, further advances are on the point of making it possible for women to be completely liberated from their biology by means of the development of artificial wombs

[. . .] Technology has now advanced to a stage at which the gender-role system is no longer necessary.'

This might have been a rather optimistic reading of the state of reproductive technologies in the early 1970s, but it was not complete fantasy: scientists had been experimenting with growing both animal and human foetuses outside female bodies for decades by this time. CHOP might like to depict its research as an unprecedented paradigm shift, but it's actually built on the back of a long and international body of scientific work. And while the CHOP team got a lot of attention when the biobag paper was published, there are scientists across the world, in Asia and Australia as well as other parts of North America, who have been working successfully with artificial wombs for years, and are racing the CHOP team to be first to try their devices out on human foetuses.

*

'This is not a new field at all,' says Matt Kemp, a little wearily. He runs the perinatal laboratory at the Women and Infants Research Foundation (WIRF) in Western Australia, and his team's artificial womb, Ex-Vivo Uterine Environment or 'EVE' therapy, reported its first great successes in a paper published a few months after the CHOP team's research. The biobag completely stole EVE's thunder, and although Matt makes little reference to the biobag, he does sound somewhat narked about it.

'There was a group out of the Karolinska Institute in Sweden that published a paper in 1958 showing the use of this sort of platform with pre-viable human foetuses,' he continues. 'There were groups in Canada in the early 1960s that were doing short-term, twelve- and twenty-four-hour experiments with sheep using this system. As early as 1963, the Japanese did the most seminal work in the field. In the 1990s the Japanese were using goats and running them out to very similar or equal – to three weeks or whatever came out of Philadelphia. More recently, there's a group in Michigan doing some work in this space. Anyone who tells you they've done this for the first time and what they're doing is novel and new is being a little disingenuous.' He doesn't name any names.

There is no patent application for EVE ('My view is that it's not patentable,' he says, exasperated. 'This has all been in the public domain in various forms since 1958') so Matt is happy to answer questions about it. I can't come to his lab in Perth because he's in Boston at the moment, studying business and leadership at Harvard Business School. We're speaking on the phone during a break between his classes.

'How come you're studying business?' I ask.

'Well, because like most other things these days, science is a business,' he says.

Today, Matt only wants to talk science. I ask him why he decided to name his artificial womb EVE, the name of the first woman and the mother of mankind, and it's obvious he doesn't want to get drawn into broad discussions about the symbolism of his work. 'It was just a convenient way of describing it, I guess.'

Matt has been developing EVE since 2013, in collaboration with a team of researchers at Tohoku University Hospital in Sendai, Japan. No official images of EVE have ever been released, but I found a YouTube video uploaded to the official WIRF channel and watched it just before I rang him up. It looked like it wasn't supposed to be online: it was clearly filmed on a phone and had had only fifty-six views in just over a year. Having only seen CHOP's promotional video and the carefully sanitized images of the lambs they submitted with their paper, this forty-four-second clip made my jaw drop.

It begins with beeping monitors in a NICU. The even, regular trace of a healthy heartbeat thumps in red across a black screen. The camera pans down to the incubator beside it, and instead of a baby there is a lamb submerged in yellowish fluid in a transparent bag. Its chest rises and falls, its nostrils flare. The camera pans again, up from the lamb's woolly abdomen to a mass of tubes protruding from the semi-open zip, like veins filled with blood. With its amateur camerawork and bright body fluids it's far more visceral than the carefully considered footage released by CHOP. And it's disturbing, uncomfortable viewing. This is what an artificial womb really looks like.

Still, the EVE therapy system looks like a biobag, and when Matt describes it, it sounds like one too. 'Extremely preterm babies are not really small babies: they are more akin to a foetus. That's the basis that we work on. We try and work with the anatomy and the physiology that they currently have, rather than trying to force them to adapt to life outside the uterus. That means making use of the umbilical cord and the foetal heart, and keeping these foetuses viable and protected under a layer of amniotic fluid, and hopefully allowing them to grow in the same way that they would otherwise.'

'You call them foetuses rather than neonates,' I say. 'Does that mean you don't consider the lambs to have been born when they are put into the system?'

'We do not.'

'So birth is opening the bag?'

'Well, I would say birth is when you've cut and occluded the umbilical cord. That's when you have agency as an individual person. My understanding is that until the cord is cut and clamped, then you are not born.'

Artificial womb technology is redefining birth: it's no longer about being pushed or pulled out into the world, it's about being separated from the life support that you rely on as a foetus. You can be separate from your mother and still be officially unborn.

Just like the vegan meat makers, Matt talks about his work like it is simple stuff – home brewing, rather than frankenscience.

'How on earth are you able to plug in through the umbilical cord?' I ask.

'It's not as tricky as you might think, once you've worked out how to do it.'

'What's in the amniotic fluid? How do you make it?'

'It's kind of like Gatorade, really. It's a salt-protein-water mix.' Just like Mike's description of the medium they grow at Finless Foods.

WIRF's collaboration with colleagues in Japan is going to give them the edge over the other teams making artificial wombs, he says. 'Our competitive advantage is that we have a pretty big Japanese biotech company working to design the hardware for the things that we use. We need to be working with people who can run things to scale and potentially run them through an FDA pipeline. We're working with a company down in Osaka called Nipro Corporation that's a world leader. It gives us a very nice system to work with.'

But the big difference between WIRF's work and CHOP's research is that Matt's team is putting far more premature lambs into EVE. The youngest lamb foetus put into the biobag was at 106 days' gestation; Matt has been working with animals as premature as ninety-five days. He's cautious about translating this into human terms, but it works out as somewhere between twenty-one and twenty-three weeks. No one else has ever reported working with foetuses this young. And while CHOP grew their lambs for several weeks and let some of them live after the experiment, Matt's team chose only to keep them in the artificial womb for a week, and then killed all of them to analyse their organs. He says they could have easily kept them alive for longer, if they had wanted to. 'These are very stable, very healthy animals at the end of their predetermined time point.'

Even in a week, the lambs change dramatically inside the artificial womb. 'They grow, absolutely. They get bigger. Lambs at this gestation will put on about forty grams a day. They flex and extend and they swallow. I've never been pregnant, but my wife has been. Her view is that a foetus does those sort of movements: it kicks, it flexes its legs, it has a wee wiggle and it goes back to sleep for a while.'

I wonder if he has feelings about his invention as a father, as well as a researcher. 'How does it feel to watch those kinds of changes if you're going in there each day?'

'It's pretty remarkable stuff. From a basic science perspective, you've built a placental knockout model.'

I try again. 'What about from a human perspective? Did you get attached to them?'

'Yeah. You certainly do get attached to these little guys. You're rooting for them.'

'Did you name them?'

'Yeah, they get named.'

'What are they called?'

'Oh, I can't remember.'

I guess if your ambition is to put the tiniest children on earth into a plastic bag, it's better not to be too overcome with paternal feelings towards them.

But clinical trials in human babies are a long way off. 'Anyone who tells you that they're going to be doing this in two years either has a wealth of data that is not in the public domain or they are being a bit sensationalist.'

'Are you talking about anyone in particular, there?'

'I am not. I am making a general comment,' he says firmly. 'All of the experiments that have been done to date are done on foetuses that come from pregnancies that are otherwise healthy and would be ongoing if they hadn't been interfered with by the research team. That's simply not the case for a twenty-one-, twenty-two-, twenty-three-week human foetus. These are not going to be healthy babies. There's a reason why they are being born preterm.' By creating a device to gestate such premature babies, both Matt's team and CHOP have set themselves a task beyond simple ectogenesis.

'The threshold barrier for getting this into clinical use is going to be incredibly difficult. If you're going to create an argument that an ethical committee will buy, you've got to have an odds-on chance of delivering an outcome that is several orders of magnitude better than the existing technology currently in use,' he says. 'What is the likely first demographic for this platform? I think we're talking about a very sick twenty-one-week foetus that essentially has a zero chance of survival on anything we have existing.'

I wasn't expecting him to say that. It completely floors me.

I lost a baby at twenty weeks – a son, who would have been my second child. There was nothing wrong with him. He was perfect. I got appendicitis when I was nearly nineteen weeks pregnant, although I didn't know it then. I spent a week in hospital while obstetricians and gynaecologists scanned and poked and took my blood as they tried to work out why I was ill and what they should be doing about it. And then I went into labour. It happens: if you are pregnant, a serious infection can make your cervix open. In between contractions, the obstetrician told me that if I had been twenty-four weeks pregnant everything would have been different, but as I was at twenty weeks I should just let nature take its course. Even though the son I gave birth to was a proper baby who was wrapped up and given to me to hold and behold, he died while I was giving birth to him. A miscarriage, not a stillbirth.

This happened three years ago. Since then, I've had my appendix out, and I've had a baby daughter: the one who guzzles cows' milk and shepherd's pie. But, like anyone who has lost a baby, I will always be haunted by the memory of the child I never had, and what could have been done differently for him. If an artificial womb might save the life of a very sick twenty-one-week-old foetus, could it not also be used to save a twenty-week-old who was perfectly healthy, but unlucky enough to be inside a woman who was ill?

I swallow hard. 'If the first time you put a human foetus in your system, it will be a foetus that is not going to be viable otherwise, can you see how questions are going to come up about pushing the boundaries of viability? Can't you imagine parents of even more premature babies wanting their child to have any chance at all that an artificial womb might give them?'

'I think this is actually a really easy question,' he replies immediately. 'This is a human – or this is a foetus, or this is a baby – that's *sick*. If you had a three-year-old that was particularly unwell and somebody was developing a new therapy for it, would you have any qualms about that?'

'Of course not.'

'So there you go. From our perspective, this is no different.'

In other words, so long as they have a chance to save a baby's life, they will try to do it. But there are limits to what they can do.

'We don't actually think we are shifting the border of viability further and further. The pragmatic reason for that is, if you can't get a catheter into it, and the heart is not sufficiently developed to drive blood through the system, then it's not going to work. So any concerns about harvesting eggs and putting them into these artificial devices are completely abrogated by that. It's just not practically possible.'

*

While partial ectogenesis is likely to be with us within a few years, it's certainly true that complete ectogenesis, from conception to live birth, is not yet practically possible. But as we get better at extending the lives of embryos outside the womb in the weeks following conception, and as we learn how to keep ever more premature babies alive, there will come a time when these two points meet – by accident, if not by design. We are getting closer to that point every year.

It used to be thought that human embryos could only be grown from conception outside the womb for a week, the time when they normally implant in the uterine lining. But in 2016, Professor Magdalena Zernicka-Goetz's team at Cambridge University succeeded in keeping human embryos alive and intact outside the human body for thirteen days by bathing them in a special medium and incubating them. With the correct cocktail of growth factors, the embryos implanted onto the bottom of the dish and early placental cells developed.

Scientists will only keep human embryos conceived though IVF alive for fourteen days because of an ethical convention of halting research before the 'primitive streak' (a row of cells that marks the beginning of what will become the brain and spinal cord) appears on day fifteen. The Cambridge team's embryos had to be killed because of this fourteen-day rule; it's likely that they could have survived many more days, if they had been allowed to try. Since 2016 there's been widespread debate about whether the limit should be extended to twenty-one or even twenty-eight days, because of the enormous scientific potential of being able to closely observe embryological development outside the human body. The fourteen-day deadline is a voluntary ethical limit only officially observed by seventeen countries. There is nothing to stop scientists in North Korea or Russia from growing human embryos for as long as they can.

In animal experiments, researchers have gone a lot further. In 2003, Dr Helen Hung-Ching Liu and her team at the Center for Reproductive Medicine and Infertility at Cornell University managed to grow a mouse embryo from conception almost to term using bioengineered womb tissue on an extrauterine scaffold. If R&D money continues to pour into the clean meat industry, our ability to culture tissue will make it even more likely that uterine tissue can be grown and used in this way.

Of course, the way an embryo develops is still a black box, and we have so much more to learn about what happens in the first and second trimesters of pregnancy. But by growing an embryo outside the human body for longer, the lid on the box begins to open. Reproductive medicine is driven by ambitious doctors and researchers, powered by a force as great as the human drive to reproduce, and funded by a customer base prepared to pay whatever it takes to fulfil that imperative. The more we understand, the more likely it is that full ectogenesis will become possible. There is too much pressure – scientific, medical but also commercial – for this not to happen. The obstacles will be ethical and legal, rather than technological.

IVF was once science fiction, then an ethical conundrum, then the cutting edge of assisted reproduction. Now it's a normal part of making families, an acronym everyone understands, uncontroversial enough to be advertised on the Tube. The right to create a baby outside of the womb is recognized by the NHS, which covers the cost of couples having a chance to conceive their own biological children this way. Things that once seemed unnatural can easily become mundane.

Once bags and tubes can replace a womb, pregnancy and birth will be fundamentally redefined. If gestation no longer has to take place inside a woman's body, it will no longer be female. Just like baby formula meant men were equally able to feed their babies, ectogenesis will mean bearing children no longer belongs to women. And the meaning of motherhood will also be changed, forever.

CHAPTER ELEVEN

Immaculate Gestation

'Pregnancy is barbaric,' Dr Anna Smajdor declares. 'If there were any disease that caused the same problems, we would regard it as a very serious disease indeed.'

I am sitting on Anna's green sofa in her office on the campus of the University of Oslo, opposite a calendar featuring photos of her cats. She is swivelling from side to side on her chair, anchored by an elbow on her desk. There's a green scrunchy on her wrist; her dark hair reaches to her chest. She is a bioethicist and Associate Professor of Practical Philosophy here, but her small frame, animated face and expressive eyes give her the air of a mischievous teenager.

'The amount of women that suffer tears and incontinence and things that damage them for the rest of their lives is really high, yet it's not adequately recognized in society,' she continues. 'This is all tied up with the strong value that we attach, not just to motherhood, but to giving birth. We expect women to joyfully go through this process. It's worth talking about, just to shine a spotlight on what we expect women to go through for the production of new citizens.'

I've been eager to meet Anna ever since I read her two groundbreaking academic papers on artificial wombs: 2007's 'The Moral Imperative for Ectogenesis' and its sequel, 2012's 'In Defence of Ectogenesis'. The first paper set out how women bear the burden of society's drive to reproduce, how 'a man can currently use his wife or partner as a surrogate to carry his child', and how the natural difference in reproductive capacity perpetuates the subjugation of women. 'Pregnancy is a condition that causes pain and suffering, and that affects only women. The fact that men do not have to go through pregnancy to have a genetically related child, whereas women do, is a natural inequality,' she wrote in the second paper. 'There is a fundamental and inexorable conflict between the demands of gestation and childbirth and the social values we share as human beings: independence, equality of opportunity, autonomy, education, and career and relationship fulfilment [. . .] Either we view women as baby carriers who must subjugate their other interests to the well-being of their children or we acknowledge that our social values and level of medical expertise are no longer compatible with "natural" reproduction.'

By anyone's reckoning, gestation remains the most significant imbalance that exists between the sexes. The division of labour in family life begins with pregnancy and continues through birth, breastfeeding and parental leave, setting up a dynamic in which the gulf between maternal and paternal input is generally vast, no matter how progressive a society or how well intentioned and determined a father might be. From the beginning, women are more expert in meeting the needs of their kids. It starts with the placenta and breast milk, and ends with packed lunches.

Anna argues that ectogenesis would allow reproductive labour to be redistributed fairly in society in every sense, so there is a moral imperative for research that advances the development of artificial wombs. Published before any biobag or EVE system existed, her papers assume 'perfect' ectogenesis could exist: an artificial womb that functions just as well as a healthy female uterus, where there would be no technical issues to make it any more dangerous than a natural womb, in a society that upholds the rights of women.

You can't blame me for assuming Anna must be a diehard feminist; she's quoting radical feminist Shulamith Firestone when she calls pregnancy 'barbaric'. But when I ask how important feminism is to her work, she baulks. 'My interest is not rooted in feminism per se. I'm interested in questions of justice and the way that human bodies are expected to produce things, or are acted upon in various ways by the state and medicine.' Ectogenesis doesn't fit into neat categories of thought, and neither does Anna.

'This is my pet subject,' she says with a playful smile. 'I've always been fascinated with reproduction, and especially pregnancy and childbirth. I think they are really *strange*. And when you look at the way in which different creatures reproduce, it's not at all a given that it has to be the way it is. I remember my mother saying to me when I didn't want to go to the doctor, "Oh, wait till you have a baby – your body belongs to everyone then." There's such an unquestioned assumption that women will have babies, and a real failure to notice how bizarre it is that we have to produce new human beings out of our own bodies. And also how risky and dangerous a process that is, even with Western medicine.'

To demonstrate her point, she tells me about the time when a colleague was having her wisdom tooth out. Anna suggested they should film it, as a beautiful experience to share and savour: 'Here it comes! And look, here's the stitching! Wow – you did that without any painkillers!' This makes me laugh out loud, both because the crude comparison to childbirth is completely perverse and also because I can kind of see her point. Our attitude to birth *is* very strange. There is blood, pain and stitching even if it all goes well, and we are meant to ignore it all. We fetishize the pregnancy and birth part of motherhood.

'We've become much more dependent on surgical interventions in pregnancy and childbirth because in the past women and babies died – that was sad, but that was how it was. These days, people survive and go on to have narrow-hipped, large-headed babies. We are actually making ourselves *more* dependent on medical intervention in childbirth. Modern childbirth is as safe as it is, very largely, because of antibiotics.' In the face of a looming antibiotic-resistant catastrophe, the future for mothers looks apocalyptic.

Rates of maternal mortality and stillbirth are going down globally, but Anna says that isn't necessarily all good news. 'It doesn't mean that you or you baby have come out unscathed. The more medicine advances, the more women do get scathed,' Anna says. 'The ways we can regulate and monitor the foetus while it's in the uterus has an impact on women's lives, what they are allowed to do, the kinds of medical interventions that they have to undergo. I don't see any great breakthroughs on the horizon in terms of maternal–foetal medicine, but I do see a trajectory towards knowing *so much* about the foetus and what's good or bad for it in the uterus that women's lives become almost as though they were ectogenetic gestators themselves. Their whole function becomes about maximizing what's good for the baby.'

I wouldn't have described it that way at the time, but I have definitely felt like an ectogenetic gestator. I have had to lie back and stare at hospital ceiling tiles, trying not to panic while a twenty-centimetre needle was plunged into my belly so that doctors could extract my son's DNA because something came up on a routine scan that made them think he *might* have a chance of *perhaps* having Down's syndrome. (He did not have

Down's syndrome, or anything else wrong with him, but then I got appendicitis.) I have had to stop myself from gagging while forcing down a cloying glucose concoction after which my blood was taken and retaken because a late scan of my daughter showed some things which *might* have indicated that I had gestational diabetes which *could* have threatened my pregnancy. (I didn't have gestational diabetes.) I have had to lie with my legs clamped apart on an operating table while a surgeon stitched up my cervix because a scan showed I was *at risk* of going into another early labour. Being pregnant is a remarkable, life-changing experience, and I loved carrying my first child, but I have never felt more like a thing than when I was receiving maternity care. Most of the time I was being acted upon there was no reason for it, other than that my very able and dedicated doctors knew too much about what *might* be going on inside me.

'In countries where abortion is legal clearly the foetus's interests are not placed above the woman, but as soon as the foetus becomes a patient – and it necessarily becomes a patient whenever the pregnant women is being monitored or treated on behalf of the foetus – there is a strong expectation that the baby's interests outweigh those of the mother,' Anna says.

'Which mothers go along with.'

'Yeah. Because it's part of showing that you are *already* a good mother. And there is almost no crime worse in our societies than being a bad mother.'

Anna is not a mother. She tells me without me asking about it. 'I don't have children and I've never wanted to have children, but I have at various times of my life been under some, er, pressure from various people to do so. One of the things that has struck me when I have been thinking about it as a possibility is, if I am pregnant – especially someone like me, who's written all this stuff about pregnancy – *everyone* knows! The whole concept of medical confidentiality is just blown out of the water,' she says. 'That very public aspect of pregnancy was disturbing to me.'

I see how the idea of being visibly pregnant would be tricky for her. I never wanted the people I worked for to know I was pregnant, and my career isn't predicated on the idea that pregnancy is barbaric.

'I do have children,' I say, 'and I didn't necessarily want everyone to know I was pregnant when I was, whereas my husband got to reveal it to whoever he wanted, whenever he wanted.'

Something changes in the room after I say this. I could be imagining it. But it feels like the personal information we have shared remains hanging in the air, drawing an invisible curtain between us. Her interest in ectogenesis is intellectual and academic; she can look at it through brutally logical eyes, and I can't.

The crux of Anna's argument is that human beings have evolved, both physically and socially, to the extent that the current way we have babies isn't working. 'There's a lot of talk about how governments and employers need to accommodate pregnancy and reproduction, but you just *can't*, because the most important years for women's careers, where they are establishing their careers, are the ones in which medics are telling them they have to have babies. There is no way of being pregnant and having a baby without it having some impact on your work life.' She seems to be assuming that the world of work and the trajectory through it is fixed and immutable, so the answer is not trying to change the workplace or the means of production, but the means of reproduction instead. It's a depressing assessment of what needs to be done to give women true equality.

We are sitting in a spotlessly clean and ordered modernist

university campus in Norway, one of the most progressive countries in the world, renowned for its generous parental leave and childcare options. This is one of the best places in the world to be a mother.

'If we made it as easy for women everywhere to have babies as it is in Norway, wouldn't so many of the inequalities that women face today disappear?'

'Maybe, but then birth rates go down,' she says simply. 'That's what's happened in Norway.'

And she's right – a few months previously, the Norwegian Prime Minister Erna Solberg made a public plea for her citizens to have more babies, fearing that current birth rates would mean the welfare state would collapse with so few young taxpayers to support it. 'Norway needs more children,' Solberg said. 'I don't think I need to tell anyone how this is done.'

'Societies with very generous provisions tend to be wealthy,' Anna continues. 'That means women have greater educational opportunities. In Norway everyone goes to university, and nearly everyone does a master's as well.' She rolls her eyes comically. 'It creates a sense of, "I have an education, I can look around, I can choose what sort of identity, what sort of career I want." Having children becomes one of many possibilities. The time at which having a child becomes the all-important goal in life, if it happens at all, doesn't happen until various other important goals have been achieved. If we don't get ectogenesis, society has a really enormous need to reinforce this maternal role of women.'

Anna was 'not very surprised' when she first saw the images of CHOP's lambs. 'I would say that those people were clever in their –' she chooses the word carefully – '*marketing*, with the image and the news that surrounded it. And of course, being unwilling to talk about ectogenesis is part of the PR sort of approach. Scientists are always very quick to say, "We're not at all interested in ectogenesis, that couldn't be further from our minds; all we are interested in is understanding gestation better and saving premature babies." That, I think, is one of the things that really concerns me about the insidious movement towards ectogenesis as a means of saving babies, which I don't think is at all likely to be beneficial for women.'

Instead of pouring resources into saving premature babies, Anna says we should be growing them in artificial wombs from the start. 'It's more likely that if we could find a whole alternative to gestation, then that would actually have better outcomes, because it's a trauma for the foetus, being removed from the uterus, even if it then goes into a biobag and manages to survive.'

'Full ectogenesis would actually be ethically preferable to the biobag?'

'Yes.'

Anna clearly loves using cold, hard logic to poke hornets' nests. She made headlines in 2013 when she wrote a paper which argued that compassion shouldn't be a necessary requirement in heath care, and that compassionate doctors and nurses could become 'ultimately dangerous' because they were more likely to burn out. But her work on ectogenesis remains her most controversial. Her parents thought it was 'horrific', she says. And they weren't the only ones.

'I got a lot of hate mail.'

'From what kinds of people?'

'All kinds of people. Men, women, feminists, men's rights activists. Conservatives and Catholics, of course, hated it.'

She tells me about the sarcastic message she got from a Vatican email address. The writer complained that he found shitting a degrading and painful process, and demanded that something was developed so digestion could take place outside the body and he would no longer be humiliated and damaged by it. (Anna replied saying he had her sympathies, but she was not an engineer and couldn't offer him practical solutions.)

Like Oron Catts, Anna uses provocative, outrageous ideas to raise difficult questions. And it works: she has made me really think about how messed up our notions of what 'normal' childbirth, pregnancy and motherhood are.

If Anna's perfect ectogenesis could ever exist, there is a long list of women who would want to use it. Women with epilepsy or bipolar disorder, for whom pregnancy would mean risking their lives by coming off medication that would damage their foetus. Women diagnosed with cancer while pregnant, who currently have to choose between saving their baby's life by continuing their pregnancy, or saving their own by starting treatment – even partial ectogenesis would be transformative for them. Tokophobics, who have experienced sexual abuse that has led to a pathological fear of pregnancy and childbirth: women who desperately want children but are unable to bear the prospect of carrying them.

Then there are women without wombs. One in every 4,500 women is born with Mayer–Rokitansky–Küster–Hauser (MRKH) syndrome, meaning their uterus hasn't developed. Others have had their wombs removed for medical reasons: survivors of uterine or cervical cancer, and women with severe and debilitating endometriosis (Lena Dunham has written about having a hysterectomy at thirty-one for that reason). These women currently qualify as possible candidates for uterus transplants. Since 2001 around forty women have had the procedure, resulting in the birth of around a dozen babies. But it requires immunosuppressant drugs and major surgery – on two healthy people, if the donor is alive (which has almost always been the case). The uterus is not a vital organ; while other transplant procedures save lives, this does not. If uterus transplants are rolled out on a wider scale, there will be even greater competition for access to transplant surgery than there is today. Artificial wombs bypass these ethical conundrums.

Ectogenesis will also help women in circumstances that are much less likely to attract public sympathy: Sahakian's current social surrogacy clients, and older women, whose bodies are less able to support a pregnancy but whose male equivalents get to have babies without a thought. Ectogenesis would mean pregnancy emancipated from age. You could conceive an embryo while you're young and grow it in a bag after you retire.

But perhaps the population most likely to be emancipated by this technology are those not born female. For the single men, gay men and trans women desperate for their own biological children, the artificial womb could be their key to reproductive equality.

*

It's six thirty on a Friday night, and London's Barbican Martini Bar is buzzing. Behind a velvet rope, beyond a sign saying, *Fertility Fest Seed Reception – By Invitation Only*, Michael Johnson-Ellis is surrounded by women in their late thirties and early forties. He's making introductions like a matchmaker, shaking hands with his right, an espresso Martini in his left.

Michael has just given a talk at Fertility Fest with his husband Wes – entitled 'Who's The Daddy?' – about all the awkward and offensive questions they get asked when people learn they've used a surrogate to become parents. Known as 'TwoDaddies', the Johnson-Ellises are bloggers from Worcestershire who promote UK surrogacy and run an online support group for intended fathers. Together since 2012 and married since 2014, they have a two-year-old daughter, Tallulah, and a son on the way, as well as Wes's older daughter from a previous straight relationship.

Michael spots me and waves me over to a quietish seat near the balcony. We sink into one of the tub chairs, and he launches into the story of the 'journey' he and Wes have taken to become parents.

'I had been in straight relationships. I got *married* at twenty,' Michael says in his gentle Brummie accent, laughing at the absurdity of the idea. 'I *know!*'

'Did you always want kids?'

'Oh *God*, yeah.' His face darkens. 'My decision to come out was either, do I stay married and commit suicide, or do I come out and sign off that I will never be a dad? Back in 2001, I knew no gay men that were being fathers, so I had written it off. I've witnessed so many men within my community hang them- selves, take tablets, and I didn't want to do that to my parents. I was at a crossroads: trading off being a dad, if it means being happy with someone I love, or being in a relationship where I'll have kids, but it won't end well.'

By the time he and Wes met, the world had changed: gay couples were beginning to have babies together. 'It was probably within the first week that I said to him, "Look, I'm going to sound like a real crazy woman, but do you want kids?"' Within a month, they had moved in together. A few weeks later, they were engaged. 'And then, about a year in, we were like, "Right. How do we create a family?"'

Wes joins us with a pink Martini in his hand. He apologizes for being late. 'Everybody wants a little piece of us tonight.'

For a couple that likes to act fast, surrogacy was painfully slow. They spent three and a half years trying to work out how to do it. 'We looked at Nepal, we looked at India, we looked at Thailand, we looked at Guadalajara . . .' Michael says.

'It was all changing as we were doing it . . .' Wes adds.

'It was all going to shit,' Michael nods. 'We started with Thailand, and then the Australians screwed that up because there was that case.' He means the Gammy case. 'Then India turned against the gays and you had to pretend you were married to your surrogate.'

'And then wasn't there an earthquake in Nepal?' Wes asks.

'Yes, and loads of embryos got killed. Then we went to Spain, to a clinic that had links to Mexico, and we were almost there. And I remember saying to the clinic manager, "How many British people have left with their child?" "Well, none yet." And I was like, "No, no, no."'

Wes felt it would be safer to have a commercial relationship with a woman overseas. 'When you decide to use a surrogate, all those natural things go through your mind, like, Will she run off with the baby? Going abroad would mitigate the risk. We would then have our baby, come back to the UK and never see that person again. The link was cut and we were back into our own world. We couldn't bump into them in Sainsbury's.'

Running out of options, Michael created a profile on the international site surrogatefinder.com, and within four weeks a British woman emailed saying she'd like to meet them. They drove up to meet her and her husband, and for once everything felt right. She ended up carrying their daughter, Tallulah, and is now pregnant with their son.

'She's part of our life now, which we never wanted and never set out to make happen,' Wes says.

'The relationship we've got with her now is not the relationship that we wanted at the beginning, but we couldn't imagine it any other way,' Michael adds.

'We wanted this very black and white transactional relationship, but actually now we're very comfortable, we already tell Tallulah who she is, how she brought her into the world.'

'Tallulah knows that she's growing her brother, and when he is big enough, he's going to come home.'

It sounds like they would have much preferred to have what Anna would call an 'ectogenetic gestator' – but human warmth got in the way, and they are pleased with how things have turned out. Of course, there would be no delicate relationships to navigate with an artificial womb, no one's goodwill to depend on, no one to bump into at the supermarket and cause an awkward scene.

Tallulah was conceived using Michael's sperm and an egg from a donor who had fair hair and blue eyes, like Wes. Wes is the biological father of the son they are expecting, who was conceived with an egg from a donor who matched Michael's darker colouring. In future same-sex parents might not have to engineer approximations of their family like this; scientists will be able to produce sperm and eggs from skin cells within a few decades. (Japanese scientists have already had a fair bit of success doing this with mouse cells, but human gametes are another matter.) Both men and women would be able to make both eggs and sperm, depending on what their relationship required.

Wes and Michael always wanted their own biological children; adoption and fostering weren't for them. They are almost apologetic when they tell me this, as if I might think they don't love children enough if they aren't prepared to take on the unknown quantity of a child given up for adoption. Heterosexual couples don't have to justify themselves like this.

Yet they have come to realize that biology isn't as important as they thought.

'I had to come to terms with Michael being the biological

father of our daughter, and I didn't know how our relationship would be. But what was very clear from the day she was born was that . . .'

Michael is welling up. 'Oh, it makes me *cry* . . .'

'It doesn't matter.' Wes is crying too now. 'It really doesn't matter.'

They both take a drink and try to compose themselves.

'Driving home from the birth,' Michael says, 'I sat in the back of the car with Tallulah, and I *cried*. No one had prepared me for that feeling. I always thought it was maternal, and it's clearly not maternal. From that moment, she imprinted on us, more than we could ever imagine, that whole parental love.'

Maybe mothers don't have the monopoly on fierce, animal love for their babies. We all sit with wet eyes for a moment. Then Michael says, 'Don't get me wrong – there are times when Tallulah is an absolute shit.'

The Johnson-Ellises have been lucky – other gay couples they know, online and in real life, have had a much tougher time. They tell me 'horror stories' about 'desperate' men who've fallen out with their surrogates, who are 'walking on eggshells' because they didn't build solid friendships with their carriers before jumping into fertility treatment. Some of the people who've used overseas surrogates have struggled with not being around during the pregnancy and feeling powerless.

'There are some surrogacy arrangements in America that we've been told of, where they say to their surrogate, "You can't leave the house after six p.m., you can't go more than twenty miles from where you live, you can't have sex for nine months, you can't drink, you need to have an organic diet." Because it's commercialized, that's what intended parents are saying in their contracts,' Michael tells me.

'And women are signing up to it because the money they are

paying is huge,' Wes adds. He still likes the idea of commercial surrogacy, though, because he says it means everyone knows where they stand.

Michael disagrees. 'I'm just not into the commercialization of a product in which the supply and the demand are massively out of kilter, and more and more people on a lower income will never be able to afford it.'

'A *product?*' I want to shout. But I don't. After all, that is what surrogacy is, if it's commercialized. A product, rather than a service; the product is the woman's womb. The impotence that customers have over this product leads to absurdly controlling behaviour set out in contracts that are terrible for women, no matter how much they are being paid.

Michael already knew about biobags before I first contacted him. There had been a buzz at Fertility Fest about the possibility of artificial wombs, Wes tells me, with one of the other speakers mentioning that men might be able to wear devices that gestated their babies one day. When I ask if there's a place for this kind of technology, their eyes brighten.

'Oh, absolutely,' Michael says.

'Absolutely,' Wes agrees.

'What would it mean to you?'

'If we fast forward twenty years to a time when this technology is available and ethically agreed, and works properly and has been properly tested, it would just give people so many more choices,' Wes says.

'And not just the gay community,' Michael adds. 'The women in the talks today – the emotion was *raw*, they were grieving something they never had. This would bring so much hope.'

But then there's the ick factor. If the clean meat industry has a steep hill to climb in terms of public acceptance, the artificial womb industry has a mountain.

'Wouldn't it be weird to watch your baby growing in a bag?' I ask.

'Yeah, for sure,' Michael says. 'You imagine a foetus in a lab, kicking away in an incubator . . . It's like something out of *The Terminator*.'

'It feels like *Alien*,' Wes corrects him.

'Because it's not natural,' Michael goes on.

'But it's also about people's perception of natural, isn't it?' says Wes.

'If something's not natural we tend to turn our nose up at it until it's explained to us. Until someone educates us that this is OK,' Michael says.

And, of course, the same goes for families with two dads.

'I genuinely think that having two same-sex parents is going to be the norm,' Wes says.

'We live in a small village. Middle England, middle class village. In Tallulah's nursery there are *two other* same-sex families,' Michael declares with pride.

'Can you imagine a Fertility Fest in the future where an artificial womb would be one of the options?' I ask.

Michael smiles. 'I would love that to be the case.'

*

'I'm a writer – that's all I am,' Juno Roche tells me. 'I say that because if you are trans people want you to be "an activist" too. I've never marched, I've never shouted, I've never carried a banner. And my pronouns – I quite like they/them. They/them fits me more, although I would never describe myself as non-binary. I just describe myself as trans. No need to put anything on the end of that.'

'You wouldn't want me to say that you are a trans woman?'

'No, just say I'm trans. I realize now, at the age of fifty-five, that gender was always the issue.'

Juno is wearing light make up – a bit of mascara that offsets their turquoise eyes – with light blonde highlights in their shoulder-length hair and gold hoop earrings. We are sitting in a quiet corner of the Quaker Friends House in Euston, and they lean on the side of the chair, friendly and conspiratorial, with their legs crossed in ripped denim jeans and spotless white trainers.

Juno has been a primary school teacher, a sex worker and a heroin addict, but they have found their calling as a writer of raw and uniquely personal pieces about the trans experience. In a moving article, published in 2016, entitled 'My Longing To Be A Mother, As A Trans Woman', Juno describes how, 'My one absolute sadness, my one absolute pain, is not being a mother.'

Back then, they were happy to be described as a trans woman. Juno had gender reassignment surgery nearly a decade ago, but rejects the idea that it has made them a woman. 'After the operation I was on this ward, and there were four of us altogether, trans people. And the others on the ward said things like, "Oh! My skin! Does your skin feel softer?" – two days after surgery.' Juno shoots me a sideways look. ' "No. I think you need to be *locked up*." '

There is a gentleness to Juno that manages to coexist with this frank directness. 'When people ask me about my genitals, I always say I've got upcycled genitals, or recycled genitals, or remade. To me it's like a work of art, and a political statement, but it's not a vagina,' they continue. 'That notion of being "real" . . . People say, "No, trans women are women." It's always people that aren't trans that say that.'

'You wouldn't say trans women are women?'

'No. Some people see themselves like that; I'm not going to mark anyone else's territory. But for me? *No.*'

Juno knows this is a minefield. The debate on whether trans women are women has been at the centre of controversy over the Gender Recognition Act in the UK, which would allow trans people to have their gender identity legally recognized without any medical evidence of transition: trans women would be women because they identify as women. It's caused uproar among some feminists who fear it would give male bodies access to private spaces intended to protect female bodies. Some trans activists have begun referring to those born female as 'womb-bearers', as if it is only the lack of a womb that makes trans women any different. Ectogenesis, of course, would mean trans women had equal access to gestation, and make this point moot.

But a female body, with the reproductive capacity that goes with it, is something Juno has yearned for their entire life.

'The earliest memory I have of anything at all is my mum being pregnant and me thinking that it looked like the most fantastic thing in the world. It was a kind of guttural, raw feeling. I told my teacher that's what I wanted to be when I grew up: I wanted to have a swollen belly full of babies.'

They were four when their mother was pregnant with their little brother, and used to press their head against her belly and listen to the baby burbling about inside. It was a home birth, and Juno got to meet their brother only moments after he was born. 'My mum looked *ridiculously* happy.'

Being a mother is about so much more than this, of course. 'Was it pregnancy and the joy of giving birth that drew you to it?' I ask.

'I think it was the thing of having the relationship. The relationship that I had with my mother was very good, very close, very loving, very nurturing and very protective. It just felt like the most wonderful, safe, secure space. It was the only thing that ever seemed to make sense in my head: that *tender* relationship. That kind of bond that a mother has, it roots you in the world. It definitely rooted my mum in purposefulness and in all the good stuff in the world.'

This moves me in a way that catches me off guard; it expresses so much of what I feel about being a mother. Here is someone who wouldn't call themselves a woman, or even use *she* and *her*, but is able to describe something so intimately female in such a deeply sincere, heartfelt way. Perhaps it says more about me than Juno, but I wasn't expecting to hear a childless trans person articulating it so well.

'I've had at least fifty years of mulling this over, and became a drug addict to try and deal with the pain of it,' Juno says, softly.

'To dull the pain of not being able to be a mother?'

'Yes. Yes. Because nothing made sense. Relationships didn't really make sense – we weren't going to make a baby. My body didn't make sense because I couldn't make a baby.'

Of course, Juno could have made a baby, but being a father was 'never an option.'

'It never crossed my mind that I could have been a father. It was bizarre to even think that I was male. I just used to think, I'm not sure why I've got this body. It always felt really detached. I couldn't interact with any kind of maleness. In a way, if I had been able to, it could have been easier.' Using a surrogate mother was also never on the cards. 'I wouldn't know how to relate to her. I wouldn't know where I would fit in, because I'm denied motherhood at source, as a trans person. I'm denied that kind of immediacy. There would have been a level of resentment that I wouldn't have wanted to have, and a detachment from the process, because this magical thing is taking place inside someone else.' Adoption and fostering were equally impossible; Juno was diagnosed with HIV in 1992 which they say rules them out. At fifty-five, they are reconciled to never having children.

'I wouldn't be the writer that I am if I'd had kids. I wouldn't be able to do the stuff that I do. You have to be realistic about it.' But this is clearly something Juno actively mourns. 'Even in our conversation today, there's been a feeling of, I can't go there.' They pull themselves back in the chair, their hands across their chest, their eyes glossy. 'It's a real, physical sadness. Not being a mother means I have to make sense of this life that doesn't make sense. That's work. Because the sadness would be overwhelming.'

Even in the face of biological reality, Juno held on to the hope that they would one day bear their own children. They tell me about five days after their reassignment surgery, when the surgeon came to inspect his work. The gauze used to pack out the new space inside Juno's 'upcycled' genitals had been removed so he could do 'the depth test'.

'He took out this single-use speculum and *pushed* it deep inside – my stitches had come undone, so it was *painful*.' We both flinch. 'Then said to me, "Oh that's the back," and then told me what depth I had. And I literally turned my head away and just cried. There's a back wall. I can't have a baby. It doesn't go anywhere.'

'But you knew that would happen,' I say, gently.

'I completely knew. But that's what I wanted. The space be-tween knowing and feeling, sometimes it's just like this –' they hold up their finger and thumb, a few millimetres apart – 'but you fall into that crevice. The flood of emotion was . . . It's a *cave*. I don't have a cervix, I don't have fallopian tubes, I don't have ovaries, I don't have a *womb*.'

Juno has heard every rumour and urban myth about the possibility of those assigned male at birth being able to one day carry a child inside them, perhaps by ectopically implanting a baby somewhere between the digestive organs, and has dismissed them all as fantastical or dangerous. 'I don't want to get hung up on the idea that they might ever be able to change *this* body into becoming *that other* body. I don't think they will.'

They had never considered ectogenesis before I got in touch. 'When you told me about it, I instantly said, "I'm not going to look it up, because it's not going to happen in my lifetime." Ever since you told me about it, my mind has drifted back to it and fantasized about it. You've made me think about something that might happen in thirty years, and I won't be here for it.'

'If it existed now, what would it mean to you?'

They pause, their eyes brimming again. 'For other people that are like me, it would mean the *world*. It would mean they could have the complete life experience. At the moment, being trans is about having maybe 60, 70 per cent of it, and accepting the big loss, the things that you can't have. I think that if it were possible, it would be so life-affirming, for me.'

'Wouldn't an artificial uterus be a bit weird? Do you think people would get over it?'

'Of course people would get over it,' they reply immediately. 'I went to the 2012 Paralympics and saw the athletes on the track. If you can get used to seeing people with prostheses brilliantly running, and not only running, but becoming heroic and sexy and desirable, and becoming the coolest people there are, then absolutely.'

If a womb that exists outside the body becomes a prosthesis for people who can't be biologically pregnant, it will offer new opportunities for different kinds of bonding, Juno says.

'Being able to go and look at something growing here, in this artificial thing, the connection is still *my* connection. *I* will go and look after it. *I* will go and sit next to it. *I* will go and look at it. *I* will go and take photographs of it growing. I will *talk* to it.' Juno is running away with the idea. 'You could create the intimacy that surrounds it. You could create the room around it. It would be in a physical space, and you could therefore kind of take ownership of that physical space. I can't own another woman's womb, or another woman's body. And there would be an immediacy there. That's what intimacy is: immediacy, a lack of barriers. The magic of being able to look in and see this thing, and know that it's yours.'

*

Before I leave Anna Smajdor, I ask her about the benefits of ectogenesis for people like Juno, Wes and Michael, which she's never written about.

'From my perspective, I am not very supportive of the right to have babies at all,' she says plainly. 'I think that to create another human being is the height of hubris.' Her eyes say she knows this is outrageous, but she says it sincerely. 'From a purely moral point of view, I think the relationship between parents and children is deeply, deeply problematic. The love children have for their parents is a kind of Stockholm syndrome: they are so dependent and they have to love their captor. There's something quite horrific about it, to my mind.'

By this point, I'm well aware of how much Anna isn't into having babies, but this is getting weird.

'I'm not saying it isn't love, I'm saying I don't think love is always as incredibly nice as people tend to assume,' she goes on. 'Because of all that, I don't support anyone's right to have a child. I support people's right not to have their body interfered with. Any further than that, I would not like to say that ectogenesis would be a good thing because it allows trans women to reproduce. My arguments in favour of it are not really about the right to reproduce.'

Perhaps sensing that she is losing me a little here, Anna leaves the world of philosophical logic for a moment.

'"The Moral Imperative for Ectogenesis" was a kind of thought experiment. I was trying to push the reasoning as far as I could, to look for the ways in which such an imperative might be argued. Assuming that we could get *perfect* ectogenesis, it *does* seem to me like a thing we should do, in a fully just society. The problem is that our societies are not fully just. And our societies are so heavily imbued with this idea that natural reproduction is beautiful and wonderful and the most amazing part of a woman's life. In a society that believes that, whether implicitly or explicitly, ectogenesis is going to be very problematic, and I think it's more likely to be used in ways that are detrimental to women generally.'

'What kind of ways?'

'When we talk about rescuing very premature babies, there's a risk that we might start to see a desire to rescue babies out of their mothers' wombs because their mother *is not fit* to have the foetus in her uterus,' she says.

If you can save a vulnerable baby from the dangers of premature birth, would you not be prepared save it from the dangerous behaviour of a reckless mother? You wouldn't need the *perfect* ectogenesis Anna imagines for that, or even full ectogenesis. You could just use a biobag.

CHAPTER TWELVE

'Finally. Women made obsolete'

It is five a.m. on a Wednesday in the city of Mobile, Alabama, and the queue outside the Mobile Metro Treatment Center snakes around the block. Middle-aged men in suits wait in line, women in waitress uniforms, weary couples holding hands. Most are in their twenties and thirties, and white, even though over half Mobile's population is black. They are here this morning, and every morning, to get the methadone they need to function. The relentless late-May Alabama sun has yet to rise, and they quietly stare at their shoes under the orange street lamps while they wait for the doors to open.

Barbara Harris has driven nine hours from North Carolina to be here. She is sixty-five and not very steady on her feet, but what she lacks in agility she makes up for with her presence, her unwavering self-assurance. She shuffles down the queue, smiling warmly to the jittery people.

'Do you know anyone using drugs who could get pregnant?' she asks each person, pressing pink business cards into their hands. *ATTENTION DRUG ADDICTS/ALCOHOLICS*, reads the red lettering on the cards, *GET BIRTH CONTROL GET $300 CASH*. In the top right corner, there is a colour photo of an impossibly small, angry-red premature baby in a NICU, engulfed by tubes, just like the babies in the CHOP promotional film.

Since she founded her non-profit, Project Prevention, in 1997, Barbara has bought the fertility of over 7,200 addicts and alcoholics. The overwhelming majority – 95 per cent – are female. Her mission, she says, is 'to reduce the number of substance-exposed births to zero', but the birth control she offers isn't condoms and pills, it's IUDs, implants and sterilization. Project Prevention doesn't perform the procedures itself, for legal reasons; instead, Barbara asks for doctors' letters to confirm the patient is on long-term or permanent birth control. The clients who choose to be sterilized get their $300 in a lump sum; the women who go for less permanent options are paid in smaller instalments so long as they can prove the contraceptive is still in place. Perhaps that's why thousands of the women she pays have opted for tubal ligation.

Barbara drives around the US recruiting new addicts in her Project Prevention-branded RV. It's covered with posed colour images – babies asleep next to trays with thick lines of coke, pregnant teens injecting themselves – under the slogan *Babies deserve to be DRUG & ALCOHOL FREE*. (The models in the photos are actually some of Barbara's ten children and baby grandchildren.) The number plate is *SENDUS$$*. Barbara tells me she gets anything up to half a million dollars in donations every year. Most of her donors are white men.

'I think if there's anything that *everybody* can agree on – the left, the right and everybody in the middle – it's that it's not OK to abuse children,' she tells me at the table inside her air-conditioned RV. Her bleached hair is pulled back into a tight ponytail; her brown eyes brim with confidence. 'That's why we have such huge financial support.'

'Having a child when you're drinking and taking drugs is child abuse?'

'Yes,' she nods. 'Well, they say don't even drink caffeine when you're pregnant, so I don't know how meth could be good for a baby.'

Barbara isn't the right-wing zealot you might expect her to be. She believes in God, but doesn't attend church regularly. She's pro-choice, but not when drug addicts use abortion instead of contraception. She's been accused of racism because she's a white woman and over 30 per cent of her clients have been people of colour, but her husband is black and her children are either black or mixed race. She's the adoptive mother of five black children, all born in quick succession to the same crack-addicted mother.

'I've seen these children first-hand. I know many people that have adopted children who have feeding tubes and breathing tubes and some that don't even make it,' she continues. 'Yes, some do survive and some will go on to be normal – I've got proof of that living in my home. But many don't. So it's a gamble. And it just depends on whether you're willing to take a gamble with the lives of innocent children.'

It is all so straightforward for Barbara. If you love kids, how can you disagree with her?

'Having money gives you a lot of power over the people you come into contact with,' I say. 'Do you feel that they're making a free choice when they're dealing with you? Is it really informed consent when they're in a chaotic situation?'

'That's between them and their doctor,' she declares. 'He has to decide whether he thinks they're able to get birth control. My thoughts are for the children. Nobody has a right to force feed any child drugs and then deliver a child that may die or may have lifelong illnesses. Nobody has that right.' She shrugs, as if she can't believe she's had to explain this to anyone.

Lots of people agree, especially here in Alabama. Since the 1950s, at least forty-five US states have prosecuted women for using drugs while pregnant. There are no laws specifically aimed at pregnant women, but states have applied existing legislation to criminalize them. Alabama's 'chemical endangerment' law was passed in 2006, targeting parents who turn their family homes into meth labs. Within months it was being applied to pregnant women ruled to be endangering their foetuses, even if they went on to give birth to healthy babies. Mothers face up to ten years in prison if their baby survives the pregnancy unharmed; if the baby dies, they are looking at a sentence of up to ninety-nine years. By 2015, 479 pregnant women in Alabama had been prosecuted under the chemical endangerment law. The most common drug they used was marijuana.

Drug screening for pregnant women has become routine – not just in Alabama, but across the US. Women in South Carolina who use drugs or alcohol from the late second trimester can be charged with felony child abuse. Under Wisconsin's Children's Code, a.k.a. the 'Cocaine Mom' law, a woman can be detained in a hospital or rehab programme against her will for the duration of her pregnancy. The foetus gets its own court-appointed lawyer; the mother doesn't.

The biobag is designed to save very sick, very vulnerable babies. It will be released into a political climate where drug abuse is child abuse, and 'very sick' is open to interpretation. The actual risks posed to foetuses of using heroin, crack, marijuana and meth during pregnancy remain unclear. Babies born to heroin-addicted mothers go through agonizing withdrawal for several weeks, but heroin is not known to cause birth defects. Prenatal cocaine exposure has not been definitively linked to any long-term effect on children's growth or intellectual development. The greatest risk babies born to drug-abusing parents face is most likely to come from growing up in a chaotic home, or being exposed to legal substances in the womb, like tobacco, alcohol and some prescription medications, which are known to cause severe birth defects. But in a culture where the drug-abuse-equals-child-abuse argument has so much traction, it's unlikely that there will be nuance in the debate once an ecto-genetic solution to the problem is a reality.

Barbara came to Mobile because someone sent her an article about a woman here who had been jailed three times for using heroin during three separate pregnancies. 'Putting these women in jail is not the solution,' Barbara says. 'They'll do their time, but nothing's going to guarantee when they get out that they don't get back on drugs and have the chance to endanger another child. That's not the solution.' Her answer is to prevent these women from being able to have babies at all. By that reckoning, ectogenesis is not the solution, either. Yet if protecting children at all costs is the bottom line, an artificial womb will always be preferable to an 'irresponsible' pregnant woman. If you can't prevent a baby from being born to an addict – and, despite Barbara's best efforts, Project Prevention has been a drop in the ocean of the number of women who are pregnant and using drugs in the US – you can at least 'rescue' it as early as possible.

It would be easy to dismiss this as American craziness, but foetal rescue – or a version of it by another name – is happening already, in countries that like to think of themselves as among the most progressive in the world, and to women that aren't even taking drugs.

In a notorious case from 2012, a pregnant Italian woman flew to England for a two-week Ryanair training course in Stansted. She had a panic attack at her hotel, and called the police, who spoke to her mother on the phone. She explained that her daughter was probably suffering the effects of not taking her medication for bipolar disorder. The police took her to a psychiatric hospital, where she was sectioned under the Mental Health Act. Five weeks later, under a Court of Protection order obtained by Mid-Essex NHS Trust, the woman was forcibly sedated and her baby was then delivered by caesarean section without her consent. Essex social services immediately took the daughter she gave birth to into care, and the mother was escorted back to Italy without her baby. When the few details that could legally be made public were reported a year later, Essex social services defended themselves by saying they were acting in the best interests of the child.

Even in supposedly liberal and enlightened Norway, the state's desire to protect babies often overrides the rights of the women who carry them. Between 2008 and 2014, the number of newborns removed from their mothers immediately after birth by the Norwegian child protection service tripled. By far the most common reason given wasn't drug or alcohol abuse but 'lack of parenting skills', a vague term that includes mothers from cultures where smacking is condoned, mothers with mental health problems, and mothers known to have had lived chaotic lifestyles in the past.

If some mothers can't be trusted to look after a newborn, will they be trusted to be pregnant when an alternative method of gestation exists? Could a mother unfit to look after her own child ever be a responsible incubator?

If the future of birth means a choice between ectogenesis and natural pregnancy, our attitude to 'natural' will change forever. It's easy to imagine a future where the kind of 'help' already offered by employers in Silicon Valley and beyond, enabling staff to freeze their eggs so they can focus on the most productive years of their careers, might include the option for employees to grow their babies in an artificial womb so they can continue working throughout gestation and birth. Using a real womb, inside the body, could ultimately become a sign of low status, of poverty, of chaotic lives, of unplanned pregnancy, or of being a borderline-dangerous earth mother, in the same way that we consider 'freebirthers' who choose to have their babies without any medical input during pregnancy or delivery. 'Natural' birth itself could become the irresponsible and reckless option.

*

The greatest existential threat faced by unborn babies today doesn't come from drugs, alcohol or women 'unfit' to be pregnant, but from unwilling mothers. Ectogenesis will be able to 'rescue' aborted foetuses: they could be transferred into an artificial womb and given to parents who want them. In the UK, abortion limits are pegged to viability outside the uterus – that's why the limit changed from twenty-eight to twenty-four weeks in 1990. Full ectogenesis would mean all foetuses – even embryos – become viable, and any unborn child could be deemed to have a right to life.

Even partial ectogenesis will turn the abortion debate on its head. We think of abortion as one choice – the decision to terminate a foetus – but it's actually two: the decision to stop carrying a baby and the decision to end that baby's life. Ectogenesis will make those choices separate and distinct, for the first time. Once a woman's body is no longer the incubator, abortion can be both pro-choice and pro-life. States will be able to allow women to choose what happens to their bodies at the same time as making it illegal to end the life of a foetus. Why should the mother alone decide that a baby should die if it can be rescued by technology?

The feminist activist and writer Soraya Chemaly was thinking about this five years before the lambs in the bags emerged live and kicking onto the world stage. In a 2012 essay for Rewire.News, she wrote that 'the tension inherent in the current debate, between the rights of the woman and the state's interest in the foetus, disappears when the woman and the foetus can be safely and immediately made independent of one another. The reproductive choices of men and women become equal and women lose the primacy now conferred on them as a result of gestation.' Her piece ended with a bleak kick in the teeth for the right to choose. 'The real dystopian future is one where we look back with nostalgia at the brief period during which *Roe vs. Wade* had its fragile relevance and impact as a high point in women's reproductive freedom.'

I catch up with Soraya on the phone from Washington DC. I begin by asking her what she thought when she first heard about the biobag, and she laughs a long, dark laugh. 'I'm quite cynical and fairly pessimistic about technology as it pertains to being truly disruptive or revolutionary. I always laugh when futurist technologists, who are still overwhelmingly men, overwhelmingly white, overwhelmingly elite, declare that their technologies are progressive and disruptive, because they produce so much of the patriarchy. They reproduce so much of the underlying inequalities of all societies. It's like trying to explain to fish what water is.'

Even with the strides Matt Kemp at WIRF and the team at CHOP have been making, Soraya takes care to say she thinks it will be a couple of generations before full ectogenesis will be a viable and widespread reproductive technology. 'It is incredibly complex and I still think it will take longer than some people believe,' she says. 'But I do think it's inevitable.' It is simply the next step in the fragmentation of motherhood. Artificial womb technology – overwhelmingly designed by men – would allow for women to be no more than gamete providers, as distinct from their gestating babies as men are.

Ultrasound images show how much female bodies are already seen as vestigial in reproductive medicine, Soraya says. 'I've been arguing for years, don't show pictures of fucking developing foetuses unless you show the entire woman's body. I understand people getting pregnant and being excited, but I'm the terrible feminist killjoy; I'm like, "Oh, that's nice, why don't we just make it *bigger*?" Ultrasound was very deliberately developed to show the foetus as though it were a planet in a void, in a vacuum, in a container, in a jar. A wallpaper of blackness around it. It completely erases the woman whose body is generative.'

I can't see full-body ultrasounds catching on, but I can see what Soraya is getting at. Flake said one of the big selling points of the biobag was that it would allow both parents to view their baby in real time, as it is disembodied from any mother. And once mothers and fathers are equally separate from their babies, they have equal rights to them, an equality that comes from women surrendering their reproductive power.

Soraya accepts that ectogenesis has the potential to free women from the burdens that currently come alongside motherhood. 'I'm really torn,' she says. 'You think, Finally, can we just be done with all of the cultural weight of having this be thought of as inherent in our natures, an inevitable primary role for all of us? And that's kind of liberating.' But Soraya is also 'an avid fan of dystopian literature, particularly feminist dystopian literature,' so she sees the dark potential for this technology to be used to disenfranchise women. Even in the most misogynistic societies, she says, women are prized for their ability to bear children, 'at least as long as there remains a chance that she'll give birth to a son.' By making reproduction equal, ectogenesis will remove the one universal power that women unequivocally have and men do not.

This makes me think about how, in the ectogenetic future,

there may be children growing up around the world who contain the genes of mothers who didn't want them to exist. They will be born at a time when there is greater access than ever to genetic parenthood, when parents like Wes and Michael, who yearn for their *own* children, have so many other technological solutions allowing them to create their families. To use Michael's unintentionally brutal language, the supply would greatly outstrip the demand. These unwanted babies may have nowhere to go. In that world, some women might seek out backstreet abortions that will end their babies' lives, rather than legal ones that would allow them to live.

It's a horrifying thought. But this is what could happen if the foetus's right to life trumps the right of a woman to refuse to become a mother.

'At the moment,' I say, 'women have a right that men don't have—'

'Which is to end a pregnancy?' Soraya interrupts.

'Which is to not become a *parent*. Because ending a pregnancy means the death of the baby, the woman can choose whether to become a parent. That's a right that men don't have. This technology would be brutally equal, wouldn't it?'

'Yes it would. And it would totally wash that away.'

'And women would lose that power that they have now.'

Soraya thinks about this for a minute. 'You are describing an interesting legal equalization that will remain unmatched in terms of cultural responsibility,' she says. 'The onus will still be on women to stop getting pregnant in the first place; women will still be the ones getting pregnant, after all. She pauses again. 'I think that's very interesting, and I think it might be a very good outcome, in terms of forcing people to come to terms with really profoundly embedded ideas about mothering.'

'In one respect that's a great thing,' I say, 'but I guess what I'm driving at is, is this a right that women want to lose?'

'If you don't need women because they are actually the only ones capable of reproducing, and you already show in the society such a disdain for women, what do you do? I don't think we have an answer to that. Ideally, we could live in a world in which we were all people, and some people choose to reproduce and other people don't, and they all have dignity and autonomy in their decisions.' People, parents, rather than mothers and fathers per se. People like Juno, like Michael and Wes. 'That's the platonic ideal of a fair distribution.' But the world we live in is far from ideal, or fair.

You don't have to be a radical feminist to accept that women's reproductive rights are already under threat, particularly in America. In May 2019, the Alabama Senate passed a bill banning abortion in almost all cases, even those involving rape or incest. None of Alabama's female senators backed the ban, but there were only four of them in the thirty-five-strong senate.

'Would ectogenesis allow men to take control of birth?' I ask.

'I think there are men who explicitly would like to take control of birth, clearly, and if they could get rid of women to do that, I don't think they would have much of a problem.'

*

It is eleven p.m. on a Friday, and I'm looking at a Reddit message board entitled 'Women completely useless now: An Artificial Womb Successfully Grew Baby Sheep – and Humans Could Be Next', created on 25 April 2017, the day CHOP's paper on the biobag was published.

'another amazing achievement of male ingenuity and creativity!' the most popular comment reads.

'Good,' another one says. 'In a decade or so I'll just contract some worthless bitch for her egg and grow my own kid in a plastic bag.'

I've been hanging out on the MGTOW subreddit, the niche online community of 'Men Going Their Own Way'. Allow me to put this specific subset of straight men with women issues in context: men's rights activists (MRAs) fight to change social values and laws they think are misandrist so that men and women can coexist on a different basis; incels wish they could exist alongside women on almost any basis; MGTOWs have decided they do not want to exist alongside women at all. They are heterosexual male separatists.

MGTOWs (pronounced *mig-tows*) believe that the world has become 'gynocentric' – concerned exclusively with the female point of view – and therefore hostile to men. They say women get all the attention on dating apps, all the spoils in the divorce courts and all the advantages when it comes to diverse recruitment strategies, while men are made to suffer harassment for child support, denied the right to stop their children being aborted, falsely accused of rape and under constant suspicion in the wake of the #MeToo movement.

The MGTOW answer is not to change the world by fighting feminism like the MRAs do, but to opt out of relationships with women altogether. The most ascetic MGTOWs go 'full monk': they choose celibacy and sometimes even vasectomies to avoid the traps they see as inherent in a life tainted by female contact. It is not a movement, it is a mode of living, as the primer on mgtow.com explains: 'It exists in the hearts and minds of the next great generation of men. The Manosphere is the Big Bang of chaotic masculine disruption that will eventually bring into existence a new personal world of freedom for those who choose to be free.'

For men who define freedom as the freedom to walk away

from women, ectogenesis is poetic payback for the diminishing role of men and masculinity in the twenty-first century. The biobag has the potential to be as much of a key to male liberation as the contraceptive pill was for the twentieth-century female liberation they lament so much. Once artificial wombs and sex robots both exist, men will be able to live with the human drives to have sex and reproduce, without having to live with women.

Reddit users can vote posts 'up' or 'down'; the more up-votes a post gets, the nearer the top of the message board it appears. This encourages a particularly inflammatory kind of speech to flourish here. But threads like the one published on 25 April 2017 are not a one-off, by any reckoning. Search for 'artificial wombs' and you will find over a hundred threads in the MGTOW subreddit alone, some of them going back to the earliest days of the platform. Comments range from the pitiful:

> *i hope this becomes reality. i'm almost 40. i REALLY want to have a child. i like kids. i have money and time, i could afford to raise a child now.*
>
> *yet as much as my desire to have a child has grown exponentially as i've gotten to middle age, my desire to touch, look at, fuck, or talk to women is at almost zero. this shit can't come fast enough. artificial womb, sex robots, vr porno, endless movies and tv, my own hobbies, my OWN money, yeah, i'll take that instead of taking care of some fat cow.*

. . . to the truly disturbing:

> *Our holy duty is prying reproduction away from women (which is not science fiction, it's very doable with our current*

*technological knowledge), and then removing women
altogether, physically. Not just relegating them into sex
slavery, not just brainwashing them and artificially insemi-
nating them in cattle pens, but getting rid of them forever.
They're the destroyers of civilizations, they're the primordial
natural carnal malevolence, literal fucking cancer in human
form and the only reason we've kept them around for this
long was the fact that we physically needed them to continue
our species/race. Once they aren't needed for reproduction,
they won't be needed for anything.*

Are these men ramping up the rhetoric to impress one another and get more up-votes, or are the misogynists Soraya imagines in her darkest dystopian dreams already planning a womanless future with ectogenesis?

I have a look at who's online and posting right now. Here's DT1726. He commented on a thread about artificial wombs recently: 'Sex dolls and artificial wombs will put women in their place for sure. Their only merits are the ability to make babies. Sex dolls will stay beautiful forever and much safer investment than a real woman. Artificial wombs will make women dispos-able just as men are. They could save our civilization,' he wrote. 'Lots of women will die, that's my conclusion.'

I log in, and pick one of Reddit's randomly generated user names: StreetSetting. Nicely gender neutral; I don't want to frighten the MGTOWs by coming over all female. I open the private chat window and send a message to DT1726.

'I'm a journalist,' I write. 'You say an artificial womb could save our civilization if done right. I'd love to hear more of your thoughts on this.'

After a few minutes, up pop the three little dots that show someone's typing.

'You can ask me anything, as long as it's not personal information,' DT 1726 responds.

'What difference do you think artificial wombs will make to human civilization?'

The replies come thick and fast.

'Women have evolved to seduce men to protect and provide for them. In a society where women have forgotten their biological roles of being the mothers and home keeper. A society in which they go around sleeping with whoever they want without restriction. A society where women became empowered because of technology plus their already heightened reproduction values. They become self-entitled princesses that look down on the builder of the civilizations they are living in,' he writes. 'When women see they don't have the monopoly on their wombs anymore, they will be hit with a cruel reality that they will be phased out if they keep up with this current mindset.'

So he's a grade-A misogynist, but he's not quite calling for wholesale femicide. He hopes that artificial wombs will put women back in their 'natural' place.

'Without the advantage of being the only ones with a womb, women can take out their eggs, fertilize them and grow them inside the artificial womb. They may be encouraged to do so if they want to pursue their career. When that happens they will have no excuses whatsoever that they are oppressed and can't compete with men.' He follows this with a blizzard of links to some scientific articles on how testosterone gives people the drive to be effective. As men have most of the testosterone, they will always do better. Women will realize there's no point in trying, so they will just retreat back into the home. Bullshit cod-scientific evolutionary psychology does well in MGTOW circles. When Charles Darwin set sail on the *Beagle*, I wonder if he had any idea that this was where his theories would end up.

'You say in your post on artificial wombs that if women are no longer necessary to make babies they will die off,' I type. 'Is that desirable?'

'When talking about survival of the fittest in human society you have to look at stupid people, mentally challenge or birth defect. society still keeps them alive. we aren't that cruel.'

'Women will still be alive, but will be as useful to society as people who are mentally challenged or have birth defects?'

'women are definitely worth more than mentally challenged or disabled people,' he says graciously. 'women are more average than men.'

'Do you think artificial wombs will mean that men need not have any contact with women at all if they choose not to?' I write. 'Do you think many men will choose not to?'

'Possibly. Although it's hard to go against the natural instinct. Not many men can go full monk without any contact with women. With lover bot and realistic AI then very likely.' But DT1726 isn't interested in a sex robot and artificial womb combo. 'I went full monk already,' he explains.

'How long have you been full monk?'

'1 year. if you factor in the time before I knew of MGTOW maybe 15 years.'

'Why did you decide to go full monk?'

'Unless a man can control his lust, he'd never be free. Definitely, it's more advantageous to have an artificial woman that you can control. I still wouldn't touch it. nothing breed mediocre men better than to make him comfortable his whole life.'

I realize that English probably isn't this guy's first language. I ask what he's prepared to tell me about who he is, and he says he's from Vietnam, he works in IT and he's twenty-eight. If he's been full monk for fifteen years, that means he's a celibate virgin. Unless something very terrible happened to him before he was thirteen.

'Do you think people are more extreme here in the threads than they would be in real life?' I ask.

'Some of them I think so, especially new comers. those who just got burnt hard.'

'Is that what brings most people here – personal experience of being burnt?'

'sadly yes.'

This would be an accurate description of smithe8, the next guy I contact (smithe8 isn't his username, it's the pseudonym he asks me to use). He's a twenty-six-year-old medical student from Chicago who's only been on Reddit for two months. His first ever post was about how his life had been ruined by 'a false, preposterous #MeToo accusation'. 'Now I have developed an absolute paranoia, which makes it nearly impossible for me to talk to women I am not related to,' he wrote then. Tonight, he's written the most up-voted comment on an artificial wombs thread posted a few hours ago. His comment was, 'Finally. Women made obsolete. It's a necessity considering how hateful females are towards masculinity nowadays.'

'Are there a lot of men who'd like to be fathers but don't want to have to have a woman in their lives?' I type into the private chat box.

He replies immediately. 'For every stuck-up feminist who thinks of men as "pigs", there is a lonely man who would love to have a child but can't, because no sane man would date a feminist (if confused, "feminist" is fully interchangeable with "misandrist"). Spoiler: that man shall choose artificial womb technology.'

'Couldn't that lonely man date a non-feminist?'

'Imply she's taken. Nowadays, men just crave a normal woman.'

'Are there not enough "normal" women?'

'Nope.'

Maybe he's worked out I'm female somehow, or maybe he's embarrassed when asked to explain what he's actually written. But things seem to change.

'In your post you said, "Finally. Women made obsolete",' I say. 'Is that what you want?'

'Absolutely no lol,' he replies. 'I honestly only sh*tpost hatefully there to radicalize them and manipulate more men into going TOW. More MGTOWs equals less competition for me :)'

'If you aren't MGTOW, how come you are posting?'

'I hope to make it grow a lot,' he says. 'Under many YouTube videos there is a MGTOW comment out of nowhere. My male friend got into it all the way. There are MGTOW sign stickers even in the toilet of the pizza hut nearby. It has potential to grow into millions. We have a weird type of fight club going on. Let them fight as I pass by into future with a great wife. Less competition.'

There is something so desperately sad about this keyboard warrior trying to radicalize his fellow burnt men into the wholesale rejection of women so that he's more likely to get laid. His 'Women made obsolete' comment on the thread is only a few hours old, and it already has 250 up-votes. I'd like to think they were all from men like him who are bluffing and posturing and don't really mean it. But as the incel mass shooters have shown, it only needs one or two people to take these comments seriously for them to have dark consequences in the real world.

'I appreciate you talking to me,' I type as I sign off.

'No problem man/madame,' he replies.

The MGTOWs might not mean all that they say when they

write posts about their desire to 'remove women altogether, physically', but they are unsettlingly fluent when they write them, even with English as a second language. These are not all brainless morons jabbing at their keyboards with one finger; they are educated people who have thought a lot about this, who devour scientific papers and news reports to feed into their twisted view of mankind. They are people who may one day be doctors, lawyers or lawmakers. Decisions about artificial wombs, and who gets to use them, may well be in their hands.

Artificial wombs will be an incredibly powerful new technology. How that power will manifest depends very largely on who is demanding that technology, who is making it, who is in control of it and who is paying for it.

Ectogenesis will free women from the uncertainty, pain and vulnerability of pregnancy and childbirth, which can be such a burden when women are living, working and competing with men who never have to go through any of it. But the equality will come from women giving up a fundamental power, in the only realm where men have always had a subordinate role. Artificial wombs might be far more beneficial for men than for women.

More than any other technology I've looked at, ectogenesis reveals the chasm between an ideal world and the real one. In a perfect world, it would liberate women and save the most vulnerable babies on earth. In the real world, women are judged and disenfranchised, prosecuted and sterilized, and despised by ever more radicalized angry men.

Once IVF became mainstream, research into treating fertility problems like blocked fallopian tubes all but stopped. Why bother, if the problem can be circumvented by assisted reproduction? In the same way, ectogenesis will mean it is even harder to justify research that makes it easier and safer for women to be pregnant and give birth without being sliced, probed and torn. And there will be even less reason to try to solve the social problems that make it so difficult for women to have babies. What's the point, if the solution is already there?

Women gain so much more than we lose by bearing our own children. We gain the immediate closeness, the intimacy that Juno craves so much. We gain the creative power of being a mother, the knowledge that our children are definitely our own, the right to choose whether to become a parent at all. Having a womb makes us vulnerable at the same time as giving us great power. How can the freedom to have babies without being pregnant be worth sacrificing any of this?

Full ectogenesis will not exist for decades, but artificial wombs are coming. We still have time to try to ensure that, when they do arrive, it's in a society that values women for more than just their reproductive capacity, and that they are put to use for the benefit of people who can't be pregnant for biological rather than social reasons. We do still have time. But maybe not enough.

PART FOUR

THE FUTURE OF DEATH

The death machines

CHAPTER THIRTEEN

DIY Death

Lesley Basset is nervous, but trying to hide it behind a welcoming smile. The people streaming into the hired Covent Garden meeting hall all look over sixty: the men are in jackets and ties; the women wear pastel cardigans and pretty scarves. Genteel enough to be mistaken for a bridge club or the audience at a classical recital, they have paid to be here because they want to learn how to kill themselves. They are pinning on plastic name badges and taking their seats in the hope that Lesley will teach them.

Lesley is the new coordinator of the UK chapter of Exit International, the grass-roots voluntary euthanasia group that makes Dignitas look meek and conservative. Where other right to die organizations campaign for terminally ill people to be able to decide when they die, Exit argues that anyone of sound mind should have the right to end their lives peacefully, at a time and place of their choosing, without the permission of any doctor or state. It's what Dr Philip Nitschke, the Australian founder and director of Exit, calls 'rational suicide'.

First founded in Australia in 1997, Exit has chapters in Canada, the US and New Zealand, as well as this new one in the UK. You don't have to be ill or even old to be an Exit member – it's officially for over-fifties, but younger people are allowed to join on a case-by-case basis. For a fee, members get information, advice and also materials they can use to end their lives. So many British members are signing up that Exit hired Lesley to open the UK office a couple of months ago.

I know Lesley would rather I wasn't here today – I'm only allowed through the door because Philip Nitschke has told her she has to let me in – so I try to keep out of her way. A volunteer with a puff of white hair is handing out teas, biscuits and suggestion forms for future meetings. She tells me she's seventy-four, and a former nurse. 'Exit and the Voluntary Euthanasia Society don't get on,' she explains as she pours my tea. 'Dignity in Dying don't like how Philip is doing it; they want to work within English law, they want judicial reform. Then there's FATE – Friends At The End – who escort people to Dignitas. They don't like Philip either.' It sounds like the Judean People's Front.

I've arrived forty-five minutes early and already fifty of the folding chairs are taken. No one can tell me for sure how many British members there are, but Exit HQ estimates there are probably about 1,000, and whenever Philip travels to the UK to give his practical suicide workshops as many as 200 people pay to come to see him. Philip is on the other side of the world today, but he's very much the most powerful person in the room. There's a trestle table with books for sale, all written by Philip. There's his autobiography, *Damned If I Do*, for £25; his first book, a philosophical treatise entitled *Killing Me Softly*, for £22; and *The Peaceful Pill Handbook*, his practical guide to different suicide methods, for £20 – although Exit recommends you take out a two-year subscription to the regularly updated online *eHandbook*, for £67.50. You can fill in a green form to order nitrogen from one of Philip's companies, a key component of one of his recommended methods. Each cylinder costs £465. That's on top of membership fees, which start at £62 a year.

But the people in the audience look like they can afford these prices. They are a noticeably homogenous group: white, middle class, equally split between male and female. They are what Philip calls 'baby-boomer types who are used to getting their own way' – retired professionals, educated and independent; lively, animated, and afraid of what a life prolonged by modern medicine might mean for them. Several are already filling in the nitrogen order forms.

I take a seat on the end of the front row. The meeting hall doubles as a dance rehearsal space and there's a large mirror at one end of the room. People try to avoid staring at their own reflections as they wait for Lesley to begin.

In an old pair of Converse, a purple checked shirt and glasses, Lesley is a reluctant master of ceremonies. She is sixty-four, a mother and grandmother, and until two months ago she made her living designing cake-decorating tools. (Her website is full of mesmerizing videos of her piping perfect rows of royal icing pearls onto fondant-covered wedding cake tiers.) When she accepted the Exit job she was supposed to be answering the phones for five hours a week, but they never stopped ringing, so it soon became four days of paid work. In practice, she's working seven days a week, her cakes neglected.

She's printed a handout from her home computer with the agenda for today's discussion. It's decorated with cartoon images of a man in a hard hat lugging a green gas canister, a Jack Russell dog in sunglasses with a Martini glass in his paw, and four multicoloured pills with arms and legs who are holding hands and dancing.

As soon as she takes the floor it's clear that no one cares about the agenda Lesley has prepared. Hands shoot up, and all anyone wants to know is where they can buy Nembutal, the barbiturate pentobarbital that has achieved almost mythical status in these circles. Almost every suicide method imaginable is either painful, or unreliable, or undignified, or prolonged, or puts innocent bystanders in danger. Nembutal is the only thing that comes anywhere near to fulfilling the fantasy of sending a person to 'sleep'. It's the solution that patients drink at Dignitas, the overdose that killed Marilyn Monroe, the injection your dog gets when it's put down, and the one-time drug of choice for death row executions, until the Danish pharmaceutical company, Lundbeck, who manufactures it, stopped supplying American prisons in 2011.

There are large quantities of Nembutal within a few miles of wherever you are now, in every vet's practice, but it's a controlled substance, illegal to sell or possess privately almost everywhere in the world. You risk a prison sentence if you are caught buying it, and every year people who have never broken the law in their lives before are arrested for Nembutal possession. In April 2016, police raided the Devon cottage of Avril Henry, an eighty-one-year-old retired academic and Exit member, after a tip-off from Interpol. They seized what they thought was her stash of Nembutal, but they only found half. Avril drank the rest a couple of days later, fearing they would return and take that too, dying earlier than she had ever intended.

A year ago, Lesley gave a glass of Nembutal to her best friend, who had been living with multiple sclerosis for twenty-seven years, and watched her die after taking it.

'We had a Plan A and a Plan B,' Lesley tells the audience. 'I wouldn't let her down. I know she would be eternally grateful, and I certainly am that I found Philip and Exit.'

Plan A worked, but it wasn't easy. Instead of being the elixir of the perfect death, Lesley says Nembutal was grimmer and slower than expected. She doesn't go into details, but it sounds like this was not a good way to go. And Lesley's life was turned upside down by assisting her friend's death.

'I wouldn't recommend it,' she says simply. 'I would recommend that you do it yourself.'

In theory, you can order Nembutal online from shady vets in Latin America, China and South-East Asia who don't ask questions, and *The Peaceful Pill eHandbook* keeps members updated on the current most promising regions. I try to imagine the ladies and gentlemen here buying bitcoin and surfing the dark web and I can't quite manage it. But many have already tried. When a woman in a pink pashmina brings up the problems she's been having with some previously amenable suppliers, there's a loud murmur of agreement. Trusted sources seem to be drying up. Nembutal isn't the solution it's supposed to be.

So Lesley talks us through Plan B: the Exit Bag. I'll spare you the precise details of what that is exactly, save to say it involves entirely legal components – a plastic bag, some tubing, a nitrogen cylinder and a few other things – and it sounds appalling.

There's one of the £465 cylinders of compressed nitrogen behind Lesley – grey, with a green diamond. It comes from Max Dog, a company Philip set up ostensibly to supply gases for people who want to brew their own beer, but a legal disclaimer on the Max Dog website says its products are only for those aged fifty and over who have never been diagnosed with a mental disorder. Max Dog regulators, which allow customers to control the flow of gas out of the canister, are sold separately. They cost £325 each.

'If you fill in the form, will it all be delivered to you?' asks a man with spectacles on a string around his neck.

'No,' Lesley says carefully. 'You have to buy all the parts separately and assemble it yourself.'

She knows very well that Exit can't provide people with a complete suicide kit, but it looks like you'll need a chemistry degree to put it together.

'Can you get any of it cheaper elsewhere?' asks another man, staring at the prices on the green leaflet.

'You can, if you want to cut off the lifeblood of Exit,' Lesley answers coolly. 'Anyone can buy the component parts in the UK, but if we don't support Exit they are going to go under. With Max Dog, you don't need a backstory.' A ripple of nods flows through the audience.

Lesley hands some bits of the kit around the audience for people to play with. It's all very jolly. They weigh up the heaviness of the metal regulator. Someone passes a bit of flaccid tubing to the woman beside him and they giggle awkwardly at each other.

As I watch them breaking the ice with their pieces of suicide equipment, all I can think is, Has it really come to this? Are people so desperate to be in control of their deaths that they are prepared to die this way – and be found this way, alone and cold, with their head inside a bag? How can this be considered a 'good' death, a 'better' way to go? The alternative option – Nembutal – means that people who would never dream of going near illegal drugs would have to become drug traffickers, shelling out hundreds of pounds into the ether in the hope that whatever they are sent, if it arrives at all, is actually what it's supposed to be, and that Interpol won't break down their door if it is. How can the desire to have a 'good' death have led to these strategies?

British people don't have the right to die. 'Self-murder' became a crime under English common law in the mid-thirteenth century, and suicide was only decriminalized in 1961. Assisting another person to end their life is still a crime, and carries a maximum penalty of fourteen years in prison. In 2015, MPs overwhelmingly rejected a bill that would have allowed people with six months or less left to live to be helped to die under the care of two doctors, even though polls show that 84 per cent of the British public want the right to die.

But across the world, the right to die is becoming enshrined in law, whether it is through voluntary euthanasia (ending someone's life to stop their suffering, at their request), assisted dying (helping to end the life of a person with only months left to live, at their request) or assisted suicide (giving someone the means to end their own life). Switzerland has allowed assisted suicide since 1942, and around 350 British people have travelled to the Dignitas clinic in Zurich to die there. Euthanasia has been legal in the Netherlands since 2001, in Belgium since 2002 and in Luxembourg since 2008, and encompasses 'unbearable' mental as well as physical suffering in those countries, meaning that alcoholics and people with severe depression are included among those legally helped to die (around 4 per cent of all deaths in the Netherlands arise from euthanasia). In North America, assisted dying became legal in Oregon in 1997, in the State of Washington in 2008, and in California and Canada in 2016.

At a time when people are living longer and not necessarily better, facing an old age where they are more likely than ever to be living with chronic, painful and debilitating conditions, dementia, loss of independence and dignity, it might feel like there is a clamour for the right to die in more affluent countries, and a domino effect that means it's going to be inevitable everywhere one day. But wherever it is legal, the right to die depends on the approval of doctors and psychiatrists. It gives the medical profession more power than ever, at a time when – from climate change to vaccinations to Brexit – ordinary people are rejecting authority and turning away from experts. Why defer to people with a few letters after their names, when you can find all you need online?

People don't join Exit for the right to die; they are seeking total control over their own deaths. Faced with the uncertain future that comes with getting old, they don't want to abdicate their self-determination to anyone else. Philip Nitschke is the only medical doctor prepared to hand over the control to them. No check-up, no terminal diagnosis required. Just a declaration of age, and a credit card.

The Exit UK inaugural chapter meeting winds up after a couple of hours, but it hasn't been long enough for many of the members. There's discussion of doing a whole-day meeting next time. 'We could bring our own lunch,' someone suggests. There's applause for Lesley at the end. She's visibly relieved that it's over: her smile is warm and broad now, and she thanks me for coming along.

A gaggle of Exit members gathers around me, eager to share stories of what brought them here. Anne is a retired academic; she has arthritis, but she's otherwise well. 'I've had a good innings and I'll be seventy-five in a couple of months,' she tells me. 'Gradually I'm being closed off, can't do this, can't do that, and I can see the trajectory of my life: I'll become more of a nuisance to everybody else, and there will be more visits to hospital, more pain and unpleasantness.'

'Do you have any experience of firearms?' a man called Brian asks me. He's a retired police officer, American-Irish, and eighty years old, although he looks little more than sixty. 'About forty years ago we had one of the lads who was an officer – he put the gun in his mouth and shot himself, but he's still alive to this day in a wheelchair.' He shudders. Guns aren't the solution for people who seek the perfect death. But I don't think bags and controlled substances necessarily are either.

Christopher, a seventy-seven-year-old retired architect, wishes he could get his hands on Nembutal. 'I'm always hoping I'll come along and they'll say, "Good news – it's available from Lidl." Or a nice gift pack from Waitrose. They never do,' he says, deadpan.

When life expectancy was lower and infant mortality higher, death was part of life; we were confronted with it on a far too regular basis. In 1945 most deaths occurred at home, but by 1980 just 17 per cent did. We can now expect to live with barely any experience of death, until we reach an age where it looms upon us. It is more frightening than it has ever been. There is a huge market for whoever can promise a painless, dignified and controlled death. So long as they can actually deliver it.

*

Philip is hard to get hold of at the moment; he is busy in court in Australia, fighting for the return of his medical licence. The Medical Board of Australia used emergency powers to suspend him after a man called Nigel Brayley attended one of his workshops in Perth and later emailed Philip directly for advice. Philip didn't know it then, but Brayley was under investigation for the possible murder of his ex-wife and the disappearance of his girlfriend. He killed himself with Chinese Nembutal before charges could be brought.

Every few years, Philip makes headlines for one reason or another. He once called for prisoners sentenced to life without parole to be given the option of killing themselves. A few years ago he announced plans for a 'Death Ship' that would take people on a cruise into international waters, where he could euthanize them outside of any legal jurisdiction. Nothing ever came of it, apart from publicity. Stories like this have earned him the nickname 'Dr Death'.

The anti-euthanasia group Care Not Killing has described him as 'an extremist and self-publicist'. Not Dead Yet, a British alliance of disabled people opposed to the right to die, say he 'is not only playing on people's emotions, but he is profiting from them'. Dignity in Dying, which campaigns in favour of assisted dying, believes Philip's workshops are 'irresponsible and potentially dangerous'.

He's always embraced the notoriety, but the business with Brayley might be one controversy too far for him. Even before he lost his licence Philip spent hardly any of his time treating patients in his GP practice – for years, he has been far too busy with Exit – but he needs his medical registration back. How can he be Dr Death if he isn't a doctor?

While I try to fix a time to talk to Philip, I get messages from David. Lots of them. David had approached me as I was leaving the Exit meeting to ask for my number because he didn't want to speak to me in front of everyone else. David isn't his real name. He doesn't want his three kids to know what he's planning with Exit. None of his friends or family know. He needs someone to speak to. 'It's been very much a solo journey,' he says.

David is fifty-five, separated, and living in Berkshire. He worked abroad for a decade but came back to the UK recently after he started having chronic digestive problems that no one has yet been able to diagnose. It doesn't seem life-threatening, but it's unpleasant enough to stop him from working.

'The thought has come up on numerous occasions of, Wouldn't it be easy – no, not easy, that's the wrong word. Why bother continuing if it ain't working?' he tells me on the phone. 'I believe everything is a choice, and I also see it as a choice that at any time in life you can go, "Hmmm, I don't want to play this game anymore; I think I'll move on." So I got very interested in the method of doing it.'

Google brought him to Exit. 'When I first heard about the bag, I was absolutely horrified,' he says, 'but when I did a little bit more research into it, it seems the most simple and straightforward.' You breathe in nitrogen, he explains – 'there's no gasping or anything' – you just pass out, and then die within a couple of minutes. With a bag over your head. Nembutal isn't for him because he doesn't like the idea of having to drink anti-emetic medicine beforehand to stop him from throwing it up, and he doesn't want to rely on the Chinese suppliers that currently corner the market. 'I have trust issues with China – you don't know what you're getting,' he says.

'There's a Nembutal purity test kit sold by Exit, which is expensive. I have to say, anything you do buy through Exit – probably reasonably so, because they do have their expenses – is pretty damned expensive,' he remarks. He's worked out that you can 'self-deliver' most of the bag kit for a fraction of what Exit charge. 'That's not to criticize. It's a business, however you look at it, but I don't for a second think they are exploiting people for profit. I guess if you want it on a plate, if you want Christmas delivered, you have to pay for it.'

It's funny David should mention Christmas. In a marketing drive of questionable taste, Exit recently initiated a Black Friday deal offering an extra six months free for new subscribers to the *eHandbook*. I've been on their mailing list ever since my first contact with Exit HQ and every few weeks there's a new email with offers, or cautionary tales of those who've gone off-message and bought things from someone Philip hasn't sanctioned. 'We have said it before and we will say it again. Online Nembutal Scammers are EVERYWHERE!' one email reads. 'If you are trying to use the open internet to buy Nembutal, you are 99.9% likely to be scammed of your money. You may be threatened & even black-mailed. *The Peaceful Pill eHandbook* is the only book to continually monitor what is happening online.' In this journey into the great unknown, only the products Philip controls appear to be endorsed as reliable and safe.

But Philip's approval is worth paying for, David thinks. 'I see Philip Nitschke as an amazing character. He's under an awful amount of pressure and I don't know what drives him, but the more of his stuff that I watch, I can't fault him.'

He pauses. 'It's really cool that I can actually talk to someone about this. I appreciate it.'

Finally there is relief in his voice. Up to now, David has sounded full of despair.

'You don't know what's making you ill, so you don't know if you have a terminal illness,' I say. 'Do you really want to be making all these preparations now?'

'To be honest, whether or not I have a terminal illness, there are days, irrespective of health, where I would quite happily have said, "It's time to check out and move on now."'

'But then there are days when you don't feel that way.'

'Yeah, of course.'

'If you had all this kit at home, would you think for a very long time before using it, or have you already done your thinking?'

'I couldn't do it right now, because it's not cleared with the kids,' he says. 'Somehow, I need to have that conversation.'

David needs to have many more conversations, with the people who love him and the doctors who treat him, rather than with Exit and with me. The answers that he is looking for are much more likely to be found in his friends and family than inside a plastic bag. But that's the only solution being offered to him at the moment.

*

I meet up with Lesley a couple of weeks later in Exit's UK office – a room on an industrial estate near her home in Kent, among corrugated iron warehouses on the River Medway. This is where Lesley runs her cake business, but it's not the bright, sugary world I was expecting. We sit at a table where decorating tools lie beside suicide manuals.

She talks me through a typical day. 'First thing in the morning, when I'm still in my jim-jams, I open my computer, because Australia will have been around for a few hours by then. Then I'll check the phone messages. We might get six or eight in a day. It doesn't sound a lot, but the return calls can be quite tricky and long.'

There are two categories of caller that are the most difficult, she says. 'There are young, depressed people. You can tell they're depressed, and you can tell they're not fifty, sixty, seventy years old. That's an absolute no. We can't.' She shuts her eyes. 'You say all the stupid things: "Have you spoken to your GP? Have you had counselling?" They don't want to hear that, yet I've got to say it. Their line is usually, "They can't help me. Help me to get Nembutal." And I can't.' She winces. 'I can't. So they get off the phone and they do something worse.'

Then there are the people calling on behalf of someone else: people who want to assist a suicide. 'We have to say, "We cannot encourage you to do this,"' Lesley says mournfully. 'It's very tough. Some of the stories of situations people are in are similar ones to my story, and I could tell them things that would help. I wish I could. But I can't.'

Her story begins in 1994, before she was in the cake business, when she worked in financial services for a woman called Sylvia Alper. Five years younger than Lesley, Sylvia was already her boss's boss, 'quite an elevated career woman, quite bossy'. Lesley had split up with her long-term partner. 'I'd got over the misery of it and started to think, This isn't half bad – you can do a lot when you're on your own. She was having a horrid time with her husband and could see that there was a different life to be had.'

When Sylvia got divorced they became best friends, going to the cinema and theatre, travelling together. 'We walked our legs off around Europe. You're looking around and then you look at each other and think, How lucky are we that we're here? Just enjoying things.' She shows me a photograph from the late nineties of the two of them on a gondola in Venice. Sylvia has a thick cascade of auburn curls, Lesley has the same cropped hair that she does now. Broad smiles illuminate their faces. 'It shouldn't have worked, because we were such different people, but it just did,' she says, with glossy eyes. 'We complemented each other.'

From the beginning of their friendship, Lesley knew Sylvia had multiple sclerosis. Sylvia didn't want the rest of the office to know in case it jeopardized her chances of getting promoted, so Lesley kept it quiet. 'When she'd lose the use of a leg or go blind in one eye or whatever and had a bit of time off I knew what it was and would go and see her. But you get over that, in the early stages of MS: you get your sight back and you get your leg back.' They both found new partners, and Sylvia moved away to Eastbourne, so they saw less of each other but kept in touch over the phone. Then Sylvia stopped getting better. Lesley's fiercely independent best friend became confined to a wheel-chair, dependent on round-the-clock care.

Sylvia had always said, when the time came, she wanted to go to Dignitas. 'She phoned me to say that she wanted me to come to lunch, that she had something important to talk about. I kind of knew what it would be. That's when she said that she wanted me to do the research. It was like we were back at work and she was giving me a project to do, and I was taking notes and saying, "Right, OK." I went off and kind of did it as an assignment.'

But they quickly ruled Dignitas out. 'By now she had to be winched from chair to bed to wheelchair. She was doubly incontinent. There was no earthly way I could get her to Switzerland.' Even if they had found a way, it was too much money. 'It was going to cost about twelve or thirteen thousand pounds,' Lesley says.

'Why so much?'

She gives me a wry smile. 'There's no reason for it to cost that much, apart from that's what they charge.' The current Dignitas brochure puts the cost at around £8,300, including doctors' fees, administration, funeral and registry office expenses, but not transport, accommodation, the compulsory Dignitas membership fees or VAT. Sylvia didn't want to spend money she could leave to her husband. Plus, he refused to take her to Dignitas anyway. 'He couldn't be the instrument of her death. So whatever we did next had to be behind his back.'

'That's quite a lot of pressure to put on you. Did you ever have any doubts?'

'Sylvia was very single-minded about everything in life. So no, there was no question when she asked that she meant it.'

My question was about whether Lesley had ever doubted that she wanted to be involved in assisting a suicide, but it doesn't even occur to her to take it this way.

Lesley found the Exit website, and learned that Philip was due to give a practical workshop in London in a few months. 'Reading about his reputation as Dr Death, it might seem grim to some, but it was perfect for me.' She went along, never letting on that she was there for someone else. She 'earwigged' on conversations around her, noting down names of possible suppliers, how much the drugs cost, how long they took to arrive. She read up on assisted suicide and what the possible consequences might be for her. She deliberately left a paper trail, so she had nothing to hide when she turned herself in (she always intended to go straight to the police; she wanted to take responsibility for her part in Sylvia's death, she felt no shame in it). She emailed a supplier and sent £400 into the unknown. Then she waited.

'In those weeks, I could barely breathe,' she says, staring at the untouched coffee in front of her. 'This is the single most important thing anyone had ever asked me to do.'

To her surprise, the package arrived. Sylvia wanted to use it immediately, begging Lesley to come down to Eastbourne as soon as possible. Sylvia's husband left them alone together. 'We talked a little about what great things we did, and wasn't it great that we did them when we could, and what a life.' She breaks off and catches her breath. 'Then I can't remember who said, "Shall we do this?" but I went into the kitchen and opened the bottle.'

Lesley held Sylvia's hand as she drank the fatal dose. From what she describes, Nembutal was far from a quick and dignified death, and Sylvia's last moments were anything but peaceful. She was retching, and her eyes, nose and mouth were streaming so much Lesley didn't know if she had taken enough to kill her. 'I have no idea how long I held her,' she says quietly. 'I don't know when she died. I tried to feel pulses, but my heart was going so much I had no idea whose pulse I was feeling.'

When she was sure Sylvia was dead, she phoned Sylvia's husband and told him to come home, and then she turned herself in to the police. Lesley describes how the ambulance and police came, how she was arrested under suspicion of assisting suicide and importing a controlled substance, how she was taken into custody for the night and made to put on a jumpsuit, and she talks in the second person. 'You're searched. They take all your clothes. If you needed to go to the loo you had to be watched by a policewoman and you couldn't wash your hands because you might wash off some evidence . . . Half of you has shut down anyway, you just go into a different place, but a little bit is thinking, Blimey, this is quite an experience.'

It took ten months for the CPS to decide not to charge Lesley. During that time, her life began to disintegrate. She says she was 'emotionally broken' and her business was 'in ruins'. Her partner was angry that she'd put them both at risk: while she was in custody the police searched their home, seized all his computers and held on to them until the charges against her were dropped; he works in IT, so his business was devastated too. 'He has gone completely to pieces,' Lesley says. It's the only time I hear a hint of regret in her voice.

When Philip came back to the UK for his next workshop Lesley went along too, even though she was still facing charges at that time. She wanted to thank him, and share her story in case it was any use to him. That's when she learned they were looking for a UK coordinator to man the phones for a couple of hours a week. She started working for Exit only a month after her case was dropped.

Lesley clearly had a lot on her plate when she decided to become the British face of rational suicide. How could she know what she was letting herself in for? Was it fair to ask her?

'Why put yourself through all of this again,' I ask, 'when you know so well what the consequences could be for you, and how devastating it has been already?'

'Because it's wrong!' she almost howls. 'It's just fucking wrong.' There's a long pause. 'It's the right thing to do, that's all I can say. It's right to help people who are stuffed. They are stuck and they're worried. In later years you shouldn't be that fearful about what's going to happen to you. It's everybody's individual right to have a say in it.'

'So you do want the law to be changed, so that people have the right to die?'

'Fuck, of course I do!'

'But you'd be out of a job.'

'Don't care. Job done. I'd retire, I'd read a book – I'm absolutely fine with that. There shouldn't be this job, and the sooner there isn't, great.'

Lesley is no tub-thumper for rational suicide. She just wanted to help her friend, and wants to prevent anyone else having to go through what she had to. She has become Philip's representative in the UK because there is nothing else to offer British people.

'There are hundreds and thousands of people who are going through this *now*, not years down the line when the law changes. They're worrying about it today,' she says. 'People like that need somewhere to turn.'

*

On the second day of Philip's medical tribunal hearing, I finally get him on the phone. It's a little intimidating to finally have an audience with Dr Death himself. It's eleven p.m. in Darwin, but he is energized, and defiant in the face of the charges against him, even though he accepts that his actions may have led to a serial killer escaping justice.

'This is a case of rational suicide,' he says deliberately. 'Brayley wasn't sick, he was forty-five, but he certainly had fairly cogent reasons, I would argue, for ending his own life. The thought that he would spend the next twenty-five years in prison led to that decision.'

'So even if he was being investigated for murder it would have been rational and you would have been comfortable with the idea of him killing himself?'

'I suppose comfortable is the right word,' he replies.

Philip tells me how he arrived at this radical libertarian view of the right to die. He discovered the world of euthanasia in 1996, when there was a nine-month window during which the Northern Territory allowed people close to death to get help from a doctor to die under the Rights of the Terminally Ill Act, which was repealed by the federal government of Australia a year later. He was in his late forties and had just qualified as a doctor then; he came to medicine late in life, after a short spell in the air force, a stint as an Aboriginal land rights activist and a few years as a Northern Territory parks and wildlife ranger.

'I heard about it on the radio and thought it was a good idea and went back to sleep again,' he says. He only got involved after there was a high-profile campaign against the new right to die, spearheaded by both doctors and the Church. 'I was annoyed, very annoyed, by the medical profession's attempt to subvert what was clearly the wish of the people. They were saying every-thing that I can't *stand* about medicine – that is, in the most patronizing way, that the doctors know what's best for you, even if you think as a member of the general public that this is a good idea. I just found that so offensive.' He made his feelings known, and people who wanted to die came knocking on his door.

'I was very much of the opinion in those early days, 1996, that it made sense that a doctor could come and see you and, if you were sick enough, provide you with the drugs to end your life. Four of my patients ended their lives. I was the only doctor to use the law, and in fact for a while I was the only doctor in the world to use legislation to effectively administer a lethal injection.' I can hear the pride swelling in his voice as he tells me this.

'Exit grew out of that, because after the overturning of that law people kept coming to see me. But then I started to see some shifts: they weren't all terminally ill; in fact, there were some people who had non-medical reasons for wanting to die, and I was challenged pretty strongly by some of these people, who said, "Why is it up to you to decide?" It's really the person who's dying's decision. That became our focus: giving people practical options, rather than sitting at the feet of politicians, begging them to change laws.'

'Are you proud of being Dr Death?'

'If you got too troubled by name calling you wouldn't do much,' he sniffs. 'It's a rare day when you don't walk down the street and someone comes up and says very nice things to you. That didn't used to happen when I was writing out prescriptions for penicillin. It's nice to be involved in an important, cutting-edge social debate. It's exciting.'

'I've been looking at how much the things you sell cost,' I say. 'The handbook is not cheap. If you want to get the nitrogen cylinder and all the other stuff through Exit, the way you recommend, it's not cheap. Are you making a profit out of all this?'

'It's not cheap, but it's not cheap to travel around the world running workshops either,' he snaps back. 'The idea that one could run the organization without that sort of financial basis is not possible. It's a not-for-profit organization. Sometimes people feel that you should never earn anything if you're associated with the issue of helping people have a peaceful death. It's almost as if the issue itself doesn't allow you to break even, let alone make a living.'

He's annoyed now that I've tainted things by questioning his motives. But when he talks about his role in helping people die, he speaks in the language of business. 'Having a local presence on the ground in the UK will make a big difference. I would expect quite significant growth. Europe, in particular the UK, is a big area of interest.'

I don't know it yet, but Philip has plans to expand the market for his work in ways no one can imagine. He has an ambitious idea that transcends the legal boundaries of any state. Something far smarter than drugs or bags. Something that doesn't require anyone's help or permission. A vehicle that will drive people to the perfect death.

CHAPTER FOURTEEN

'The Elon Musk of suicide'

At least thirteen other doctors have earned the epithet Dr Death, including Harold Shipman and Joseph Mengele. Philip isn't even the original Dr Death of euthanasia, nor the most famous. That honour goes to Jack Kervorkian, the Michigan pathologist who campaigned for the organs of death row inmates to be harvested, who pioneered the use of blood transfusions from deceased corpses, and who personally assisted the deaths of 130 Americans during the 1990s.

Kervorkian invited his patients into the back of his 1968 Volkswagen Vanagon, a campervan with some of the seats removed, where he hooked them up to one of his purpose-built death machines. His first device was called the Thanatron (after Thanatos, the embodiment of death in Greek mythology), and was made of whatever he had to hand: car parts, magnets, pulley chains, coils and toy parts. It was little more than three bottles hanging on a crude metal frame, connected to a single IV line, with a large, red button on the boxed base of the device – the kind you might find on an old arcade machine. You could easily mistake the entire thing for a macabre school science project.

When Kervorkian connected his patients to the machine they initially received a harmless intravenous saline solution, but when they pushed that red button the saline would stop, and a fast-acting barbiturate anaesthetic would be dispensed, putting them into a deep coma. After sixty seconds, a lethal dose of potassium chloride would be administered, which stopped their heart. They would die of a heart attack while they slept.

The Thanatron was used for the first time in 1990. The patient was Janet Adkins, a fifty-four-year-old schoolteacher from Portland, Oregon, living with the early stages of Alzheimer's. She met Kervorkian for the first time only the weekend before her death; he decided she had the mental capacity to understand what she was doing, and he drove her to a local park the following Monday afternoon, where she died in the back of his van. Kevorkian told the *New York Times* two days later that, just before she died, 'she looked at me with grateful eyes and said, "Thank you, thank you, thank you."'

The Thanatron was a pretty basic way for Kervorkian to try to absolve himself of accountability: his patients were the instigators of their own death, because if they didn't push the button they would remain alive on the saline drip he had hooked them up to. But the Michigan Medical Board didn't see it that way, and revoked Kervorkian's medical licence after he used the Thanatron a second time. This meant he no longer had legal access to the substances he needed for it to work. His death machine of choice then became the Mercitron – effectively a gas mask connected to a tank of nitrogen and carbon monoxide, with a clothes peg stopping the flow of gas into the mask. The patient removed the peg, and precipitated his or her own death, with Kervorkian standing by.

The deaths caused uproar and much hand-wringing in America. Michigan had no law against assisted suicide at the time of Janet Adkins' death, so there was nothing to charge Kervorkian with, although attempts were made to bring murder charges against him. Most of his patients were not terminally ill, and autopsies showed that at least five were in good physical health at the time of their death. The thing that made Kervorkian so elusive was that it was his machines that killed his patients. Their deaths became depersonalized, no one's responsibility. With that came the promise of a clean, controlled death, even if the mechanism used to deliver it was messy, and the decision to seek it often chaotic.

Kervorkian's downfall only came when he left his machines at home. In 1999, he directly administered a lethal injection to Thomas Youk, a fifty-two-year-old in the final stages of motor neurone disease. Kervorkian was getting cocky: he videoed Youk's last moments and can be heard on the tape daring the authorities to stop him from euthanizing again. They rose to the challenge and charged him with second-degree murder, and he served eight years of a ten- to twenty-five-year jail sentence when he was in his seventies. He developed liver cancer and died from a blood clot in 2011, aged eighty-three – in hospital, surrounded by doctors, without the help of any death machine.

To his supporters, Kervorkian was a hero and a renaissance man. He played jazz flute and organ, and released an album of his instrumental compositions in 1997, entitled *A Very Still Life*. He painted in garish oils, depicting everything from Johann Sebastian Bach to gruesome decapitated heads streaming with blood, and gave his paintings titles like *Coma, Fever, Nausea* and *Paralysis*. (Some of his canvases were auctioned after his death, with an asking price of $45,000.) He walked the red carpet with Al Pacino, who won an Emmy and a Golden Globe for his portrayal of Kervorkian in 2010's *You Don't Know Jack*. He was an attention-seeker who got the notoriety he craved.

Not content with being 'the other Dr Death', Philip desires an even greater legacy. By entering the field only a few years later, he has an advantage Kervorkian could only have dreamed of: instead of springs, clips and clothes pegs, Philip has computers.

*

Are you certain you understand that if you proceed and press the Yes button on the next screen you will die?

The words on the blue-tinged screen are centred to hover over two virtual, clickable buttons: *No* to the left, *Yes* to the right.

Click *Yes*, and you are taken to another screen:

In 15 seconds you will be given a lethal injection . . . Press Yes to proceed.

Click *Yes*, and after fifteen seconds there is a rhythmic pumping sound. The screen goes black, except for one word:

Exit

This is the last word Bob Dent, Janet Mills, Bill W. and Valerie P. ever read. When they clicked the final *Yes* button, a lethal dose of Nembutal was delivered into their veins. They were the four people Philip helped to die in 1996 and 1997, in the nine months assisted suicide was legal for terminally ill people in the Northern Territory. Their lives were ended by Deliverance, a machine Philip invented and built, which is now in the collection of London's Science Museum.

The screen belongs to the grey Toshiba laptop Philip also used to check his emails and surf the internet. It's battered and grubby, already three years old in 1996. It was wired up to a small, hard-shelled, plastic suitcase lined with insulating foam. Inside the case lay a tangle of red and black wires, transparent tubes, valves, pumps, a pressure gauge and several syringes, including one large one which was connected to the very long, very sharp needle that Philip put into his patients.

Deliverance was actually the name of the software Philip had written, which he described at the time as 'a program for subject-controlled medically-assisted suicide', but he came to refer to the entire device as the Deliverance machine. The Rights of the Terminally Ill Act would have allowed him to administer Nembutal directly himself but, perhaps with Kervorkian firmly in his mind, he chose to design an eye-catching contraption to do it instead.

Philip held a press conference shortly after he used it for the first time on 22 September 1996. His patient – Bob – was sixty-six and terminally ill with prostate cancer. 'We shared a meal and we shared a drink, and then he indicated that he wanted to proceed,' Philip told the assembled journalists. Then he read out a statement from Bob: 'My own pain is made worse by watching my wife suffering as she cares for me, bathing me, drying me, cleaning up after my accidents in the middle of the night and watching my life fade away.' Bob's death wasn't just about Bob; it was about the burden Bob had become as he lost control of himself.

The other deaths followed quickly. Janet, fifty-two, had a rare and disfiguring form of skin cancer and had been given nine months to live. Bill was sixty-nine and had terminal stomach cancer. Valerie, seventy, had breast cancer; her death was Philip's last legal assisted suicide, and his most controversial – by her own admission, Valerie had had good palliative care and was suffering 'no symptoms', but he helped her die anyway.

Philip has posted an interview of himself on his Vimeo page, filmed a few years after the law was overturned. He sits at his desk in a light blue Hawaiian shirt festooned with bright palm trees, unbuttoned to reveal a smattering of greying chest hair. He is reminiscing about his time using Deliverance, in front of a wall plastered with newspaper headlines about him.

'I felt the responsibility weighing pretty heavily on my shoulders,' he says. 'I would go around there, had my little case, had the machine, you couldn't just forget something and say you had to go home, or "Can we do it tomorrow?" or something. People had decided that was the day they were going to die. I had to, in a sense, make that come true. I had to make it *possible*, make it *work*. And the expectation I found almost crippling.'

Philip didn't relish assisting suicide the way Kervorkian seemed to. He did not want the responsibility of making sure it all worked when the time came. Using a computer on his patient's lap, instead of a syringe between his own fingers, allowed him some kind of distance from the act he was committing, but it wasn't enough. Lesley's words from the Exit meeting echo in my head: *I wouldn't recommend it. I would recommend that you do it yourself.*

Philip's next inventions allowed Exit members to do just that. Launched in December 2002, the CoGen machine was a carbon monoxide generator consisting of a canister, an intravenous drip bag and nasal prongs to inhale the gas. Strong but commonly available acids were combined within the canister to produce carbon monoxide, killing whoever inhaled it in only one or two breaths, Philip promised. At Exit meetings, Philip swore anyone could make it using a Vegemite jar and materials you could buy legally for around $50. 'It's not rocket science,' he told the *Sydney Morning Herald* at the time. 'Anyone who has done high school chemistry can build one of these machines.' But no one is ever reported to have died using the CoGen. Messing about with strong acids is dangerous. Carbon monoxide is a poison, and anyone planning to kill themselves this way might easily kill whoever found their body, too.

When the CoGen failed to take off, Philip developed the infamous Exit Bag, which was supposed to require even less scientific wherewithal, and uses oxygen deprivation rather than poison to kill you. But there cannot be an ick factor greater than the one generated by the idea of spending your final moments suffocating inside a plastic bag. Philip knew even then that the Exit Bag made people squirm. Neither of these devices could top the Deliverance machine, with its high-tech allure, its neatness, its stability. Software seemed to confer a certain dignity on proceedings that simple chemistry and mechanics could not.

*

In July 2015, eight months after I met Lesley at the Exit meeting in Covent Garden, Philip emails me to say he's coming to London. We finally meet in the chic Airbnb he is renting in Hackney. Sumptuous oil paintings in gold frames cover the walls; there are white wooden shutters on the windows and whitewashed floorboards. He's in green shorts and another of his trademark summer shirts, brash against the impeccably tasteful white sofa.

His wife, Fiona, is trying to keep their beloved and overweight Jack Russell, Henny Penny, from disturbing us, but I'm feeling unsettled anyway. My mind races with thoughts of all the people who have died because of the man with bare knees next to me. There is no way he could quantify them, even if he wanted to. And there is something mercurial about Philip that is even more striking in person, an aloofness that makes me feel like I have to get every answer I can from him during these few moments I have in his presence, as if he's going to disappear into the ether soon, or decide he doesn't want to talk to me again.

Plus, Philip has strange reasons for being in the UK this time. He's gearing up for a one-man stand-up comedy show at the Edinburgh Fringe Festival. He's calling it *Dicing with Dr Death*. He's bursting to tell me all about it.

'Twenty days in a row, with one night off, running from six to seven, at a very nice venue called The Caves – home, it turns out, of the notorious killers Burke and Hare, the body snatchers who used to feed corpses to the Edinburgh medical school,' he says, like a carnival barker. 'A nice nexus there between crime, death and medical schools, which I will certainly be drawing upon.'

I hadn't really pegged Philip as a comedian. He definitely knows how to put on a show: his workshops and press conferences all seem to have been performances, to an extent, and yes, comedy resides in the darkest of places. But Philip? Funny? I'm not sure. There are practical reasons for this career shift, of course: Philip's medical licence is still suspended. Exit members have donated $250,000 to his legal fund, but the case remains ongoing.

He isn't bothered. 'It's an indication of authority. If you're getting information out which is so accurate that the state decides to deregister you, people will know that's good information.'

'So it's given you clout?'

'It has given me *status.*'

The comedy show will be a way of giving suicide advice, he tells me, in a climate probably too hot for him to hold his usual annual London workshop. Audience members will sign some kind of disclaimer before the show starts, but Philip has no way of checking they really are of sound mind.

His act will have an unforgettable centrepiece. It's called Destiny. 'After much research and development over the years, we've finally got a machine which will allow a person to take their own life quite easily,' he enthuses. 'I will show the audience that this is the way of the future.'

Destiny is set up on a table to our left. Philip has been calling it 'Son of Deliverance' on Twitter, but it's more like the love child of Deliverance and the Mercitron: Philip developed it after discussions with Neal Nicol, Kervorkian's long-time friend and associate, and it uses the same compressed carbon monoxide/ nitrogen mixture used in the Mercitron. Destiny consists of a familiar hard-shelled, plastic suitcase lined with insulating foam, containing a small black Raspberry Pi microprocessor, connected to a canister of Max Dog-branded gas and some nasal prongs. The microprocessor can be operated using a smartphone app or any HDMI screen, and it asks identical questions to the Deliverance software (the words 'lethal injec- tion' are replaced by 'lethal gas'). There's also a finger cuff to measure the heart rate and oxygen saturation of the person using it; when these both drop to zero, the microprocessor turns the gas off. The prototype has been paid for out of tar- geted donations from Exit members keen to try the device themselves, Philip tells me. The death machine has truly entered the crowdfunding/smartphone age.

'A member of the audience will come up and try the machine – not using the gas that the real machine will use, using an innocent enough gas – but they will see all of the pro- cess. When they press that button, they will feel that the gas will start to flow, and that their heart rate starts to falter. It will be interesting.'

Philip says Destiny will be available to Exit members and *The Peaceful Pill eHandbook* subscribers for £200 once his Edin- burgh run is finished. All the component parts are legal, but they will have to be bought separately: the app and micropro- cessor from Exit, the nitrogen from Max Dog, and the nasal prongs from anywhere you like (you can get a set for just over a quid on Amazon). Just like the Exit Bag, assembly looks likely to be a costly and bewildering process, but one with enough legal loopholes to protect the man who designed it.

'The law is flat out trying to keep up with what's happening with technology. It's like trying to shut the stable door after the horse has well and truly bolted. It may well be that those much talked about changes to legislation will come in. But it won't affect the growth of Exit.'

When the Edinburgh reviews come out a few weeks later, they are mixed. The *Daily Telegraph* gives it one star. 'Witlessly infantile,' its critic says. 'The most lamentable slab of self-publicity masquerading as a bona fide show.' This doesn't stop Philip taking an 'Australianized' version of the show to the Melbourne Comedy Festival. The *Sydney Morning Herald* critic likes it a little bit more, giving it two and a half stars. 'Laughs were sparing,' he writes.

It is not enough for Philip to give up his day job, but he does that anyway. When the Medical Board of Australia announces it will remove his suspension to practice, Philip calls a press conference and sets fire to his newly reinstated medical licence in front of the assembled cameras. 'Today, and with considerable sadness, I announce the end of that twenty-five-year medical career,' he declares. Within a few months he has left Australia for good, for a new life in the Netherlands.

*

It is four years before I see Philip again. My messages remain unanswered, my calls ignored. But I'm still on the Exit mailing list, so every few weeks I get an email warning me about suspect Nembutal bought from unapproved sources, unfair fees at Dignitas, how progressive the Netherlands is compared to Australia, and forthcoming Exit meetings. Lesley has been replaced as Exit's UK coordinator and appears to have fallen off the radar. So has the Destiny machine: after all the fanfare and press coverage that accompanied its Edinburgh debut there is little mention of Destiny afterwards, and certainly no invitation for members to purchase it.

But then an email arrives calling for proposals to be delivered at a conference Philip is convening in Toronto. It's called NuTech, and it will 'bring together experts from around the world to discuss new technological initiatives to make easier a peaceful elective DIY death'. NuTech is nothing new – it was founded in 1999 by Philip and the euthanasia campaigners Derek Humphry, Rob Neils and John Hofsess, and has taken place every few years ever since – but it has always been an invitation-only event: you have to be a right to die advocate, doctor, pharmacist or engineer to attend. This year is the first time parts of the conference will be live-streamed on the internet. And, also for the first time, there will be a competition to find the very best death machine. 'A $5,000 cash prize has been established – made possible by a generous bequest to Exit International – to an innovative proposal that advances the use of technology in a DIY peaceful, reliable solution,' the email reads.

Over the coming months, details begin to emerge about the proposals they'll discuss at NuTech. There's a monstrous-looking contraption called the ReBreather-DeBreather, designed by an American team, which is a padded mask connected to corrugated tubes that go into a blue wheelie suitcase. There's the equally ugly Australian GULPS Monoxide Generator – a small oxygen mask connected to a jerry can and some jars containing formic and sulphuric acid. (It's clearly inspired by the CoGen, and comes with the same problems associated with carbon monoxide poisoning and strong acids.) There's even a 'euthanasia rollercoaster', designed by the Lithuanian engineer and artist Julijonas Urbonas, which would kill its passengers 'with elegance and euphoria' by exposing them to extreme G-force for one minute, over seven loop-the-loops.

Then something lands in my inbox a week before the Toronto conference, and I finally understand what Philip has been up to in the Netherlands, and why he suddenly wants to open NuTech up to the public. It's a press release, entitled 'Canadian launch of world-first 3D Printed Euthanasia Machine'. Philip has a new device to unveil. He's calling it Sarco. And it makes every death machine ever invented to date look like a joke.

'Developed in the Netherlands by Exit Director Dr Philip Nitschke and Engineer Alexander Bannink, the machine was designed so that it can be 3D printed and assembled in any location,' it reads. 'On reclining in the capsule, activation uses liquid nitrogen to rapidly drop the oxygen level, and a peaceful death will result in just a few minutes. The capsule can then be detached from the Sarco machine and used as a coffin.' Sarco is a sarcophagus: the coffin that will kill you.

There are some concept pictures of a pearly-white Sarco on an empty beach, angled towards the sunrise, bathed in golden rays. This is no Heath Robinson or Rube Goldberg machine, cobbled together from spare parts. Sarco looks like a vehicle worthy of James Bond or Batman, a spaceship that will transport its user into the next dimension. The capsule is long, curved and opalescent like a mussel shell, tilted and slightly asymmetrical, with a brown-tinted transparent window. Sarco is glamorous. In the next Exit newsletter, Philip says it promises 'a peaceful, even euphoric death' with 'style and elegance'.

If the Deliverance and Thanatron machines separated death from the person who was assisting it, Sarco is the device that does away with assisted suicide altogether. If you download a death machine and kill yourself with it, how accountable can anyone else really be? Philip won't have to ship anything at all. He will be completely removed from the people who use his invention. As he writes in the Exit newsletter, 'No need to break the law. No need to import hard to get drugs over the internet. No doctor required.'

But it's more than that. No more needles, tubes and wires. No more plastic bags on heads. No more yuck factor. Sarco is the answer that rational suicide advocates have always dreamed of, and it's coming soon to a 3D printer near you, with the plans free of charge – to paying Exit members and *eHandbook* subscribers, of course. The perfect death, delivered – to anyone with an internet connection.

On the day of the conference, Philip appears on the livestream with a 1:7 3D-printed model of Sarco, which looks like it could be one of my kids' Octonaut toys. He explains that liquid nitrogen will make the machine silent – the gas won't roar like it does when it comes out of a canister – but it will also make the temperature inside Sarco drop, so users will need to dress accordingly. Apart from the nitrogen, there is one other element that can't yet be 3D printed: the digital keypad used to unlock Sarco's door. Users will get the code they need to access it (valid for twenty-four hours) only if they pass some kind of psychiatric test to determine that they are of sound mind. But Philip explains that even the keypad will be 3D printable in future. You can already print copper and electronic circuitry. It's just a matter of time.

I'm cynical enough to think that Philip has created a competition just so he can win the prize money, but it turns out that's not the case. Sarco is ineligible because it's Philip's baby. In the end, the ReBreather-DeBreather and the GULPS monoxide generator win, but they don't make it into the international coverage of NuTech. Sarco is all anyone wants to talk about, and it's breaking news everywhere from the *Sun* to Fox News to *Vice*. *Newsweek* is particularly impressed. 'Meet the Elon Musk of Assisted Suicide', its headline reads. 'His latest death machine, the Sarco, is his Tesla,' it continues. 'The Sarco is sleek – and, Nitschke stresses, luxurious [. . .] It is, in short, the Model S of death machines.'

Philip cannot get enough of this comparison. He puts it in the next Exit newsletter, and his Wikipedia page is quickly updated with his new moniker. Who cares if there are thirteen other Dr Deaths? He's the only Elon Musk of suicide.

For the next year and a half, almost all the messages I'm sent from Exit are about Sarco: how the 3D printer in Haarlem is buzzing away to produce the first full-sized prototype; how YouTube 'Sinks to New Depths in Censorship' because it removed the livestream video of Sarco at NuTech from Philip's channel; how Philip will be at the Amsterdam Funeral Fair with a virtual reality headset, so users can experience a Sarco death without actually dying.

Finally, the news I have been waiting for arrives. 'After three years in development, the world's first 3D-printed euthanasia capsule will go on display at the Palazzo Michiel at Venice Design,' the press release reads. 'I am extremely pleased that Sarco is here in the centre of the art world in Venice,' Philip writes. 'This year's Biennale's tagline "May You Live in Interesting Times" could not be more perfect.'

It's as if Philip's creation were on display at the Biennale itself. It's not. The Venice Design Fair is timed to coincide with the prestigious contemporary art exhibition, but it's entirely separate – a fringe event, if you will. Still, after Edinburgh, perhaps Philip is determined to nail all of the world's great festivals.

Kervorkian had jazz flute and oil paintings; Philip has comedy and eye-catching Dutch design.

The Venice Design Fair is free and open to the public. There will be a big press launch on the opening night, where Sarco will finally be unveiled. This, I cannot miss.

*

The Palazzo Michiel del Brusà is a Venetian fantasy of baroque majesty and exposed brickwork, right on the Grand Canal. The ground floor hall is level with the water, illuminated by afternoon sunlight that streams in through the arched doorways. A pyramid of fruit has been placed on a plinth at the centre of the room, demanding to be Instagrammed. People buzz around it in too-short trousers, long coats and ochre satin shoes – ridiculous, to my unsophisticated eyes – selfie sticks aloft. In their free hands, they hold glasses of Prosecco, or little plates with shavings of parmesan and cubes of ham.

I follow a woman in silver stilettos and a floor-length ivory cape up a flight of stone stairs. There is an enormous yellow sponge on a wooden platform. The placard on the wall says it's called *XXXXXL Sponge*, from a Dutch designer's SPONGE series, and is 'a design reflection on the damages caused by human on nature'. A doorway is covered in different-sized rubber orbs in shades of cream and grey, made by an Egyptian jewellery designer; it's impossible to walk beneath them without reaching up to squish them. There are all sorts of different kinds of mirrors and chairs, loungers and pouffes, as if this is a fair for people who like to look at their reflection and then rest. The chatter is in French, English, Russian and Chinese as well as Italian. Most of the guests only view the exhibits through their phone screens.

I turn a corner and come to a doorway. *THIS ROOM MAY*

CONTAIN SENSITIVE CONTENT FOR SOME VIEWERS, says an enticing notice. In the centre of the space, underneath angled spotlights, is Sarco itself, in Exit's trademark purple, lacquered and sparkling, dramatic and striking and very weird. The upholstered seating inside it is as elegant and reclined as any of the other chaises longues on display here. But there's a roughness to Sarco's body that I wasn't expecting: the lamination of the 3D printing is clearly visible on the grey parts of its frame, giving it an unfinished, home-made look. This is intentional, a placard explains: it has been 'deliberately left untreated in order to demonstrate the raw 3D print process'. But I'd been anticipating something more perfect. James Bond would not die in this.

And he wouldn't fit in this, either. It's small. It's definitely for the shorter suicidal person, and even then it would be quite a claustrophobic death. It might have a canopy door like the DeLorean from *Back to the Future*, but it would be impossible for an older person or anyone with mobility problems to climb through it. Could any of the people I met in Covent Garden really print this and put it together, even if they could squeeze themselves in? Would it even work, if they did? The illuminated digital entry keypad is in a little recess next to the door, but nothing happens when I press the numbers. There's a drawer at the base of the capsule where the liquid nitrogen is supposed to go, but it's fused shut. This doesn't look like a functional machine.

I follow the sound of live lounge jazz and head back downstairs, trying to find Philip. I look on the decking beside the canal, crowded with people taking more selfies. Someone has even brought a dog in a pram with them. A fat Jack Russell. Henny Penny! And there is Fiona, and Philip. The Hawaiian shirts are gone: Philip's in a beige linen jacket, a fetching straw hat and a black neckerchief. His eyes are startled to see me behind his circular glasses, but he shakes my hand. He shuffles back up the stone stairs with me to the Sarco room, his bottle of Italian beer still in his hand.

I cut to the chase. 'Does it work, this version I'm looking at now?'

'We've measured what happens to the oxygen level inside the capsule.'

'You've tested it?'

'Yeah, it works extremely well. You start at 21 per cent oxygen, which we're all breathing here, and within a minute you're down to less than 1 per cent. We sort of know what happens when you're put into a 1 per cent oxygen environment: it is actually quite soporific, disorientating, almost intoxicating. Here's Alex.'

He gestures over to a tall man in a neatly pressed blue suit: Alexander Bannink, the Dutch engineer who usually designs buses, trains, medical splints and prosthetics, but is now making Philip's ideas about death stylish for the first time. They give each other brotherly pats on the back.

'What do you think?' Alex asks me, immediately.

I don't know how to answer this. It looks like nothing else I have seen before, but it doesn't look like it works. The keypad seems like an afterthought, when it should be the first thing you work on if you are serious about rational suicide being rational. I am impressed and underwhelmed, intrigued and disturbed.

'That is a good question,' I reply. 'I think it looks like a vehicle, doesn't it?'

This seems to be the right answer. 'That was Alex's idea! To get the idea of movement. In fact, a lot of the ideas about the whole thing were Alex's.'

'How would you describe Sarco? What is it?' I ask.

'It's the demedicalization of the dying process,' Philip says, as people waft around his creation, taking photos. 'What I'm worried about, in the general trend of people seizing control of their end-of-life options, is the increasing medicalization of the process. We *aren't* really gaining control, we are *divesting* control to the authority of some other body, usually the medical profession. Sarco allows a person to say, "I make the decision, and I don't need any other 'expert' help."' Philip is the doctor gone rogue to give people true power over death.

'The only medical involvement will be initially to determine whether you've got mental capacity. Part two of this process is the development of an artificial intelligence test for mental capacity,' he continues. 'The keypad won't work unless you've passed the test. There's a lot of work going into it. And of course there's a lot of opposition, people saying it can't be done, there's no way artificial intelligence can replace a psychiatrist. It's not hard to do. Whether or not we accept it is the question. Within the medical profession, there's a lot of resistance to any form of artificial intelligence taking over their roles. There's big changes going on, in terms of what's possible.'

Alex is very proud of Sarco's green credentials. 3D printing means there's no carbon involved in transporting it. 'The base is biodegradable plastic, PLA, basically potato starch, or sugar beet starch,' he says, as if this were made of old chips, instead of a substance that actually takes decades to degrade properly. 'All the finishes are as environmentally friendly as possible, and the lacquer is water-based car lacquer.'

'Why was that important?'

'Well, because maybe you're buried in it.'

'And even if we don't bury it, we want to be environmentally friendly,' Philip interjects. 'We want to make a small global footprint. Some have come to us saying, "I want to die now because

I'm consuming resources. I've come to my natural end of life and I don't want to be a burden on the planet, I want to do the right thing by the planet." That's an increasing thing we're seeing.' This makes me think of Bob Dent, the first patient to use the Deliverance machine, who so hated being a burden on his wife. No one wants to be a burden.

No matter what Philip has said, I still don't believe that this thing in front of me works. So I ask Alex.

'It's still a concept,' Alex replies carefully. 'Because of the time schedule for Venice, the base is not functional, but the top part is.'

'Have you ever lain down inside it?'

'No, I haven't,' Philip says, taking a sip of beer.

'I'm scared,' Alex laughs.

'The arse might drop out of it. We didn't want to do that, days before the launch.'

'Would a tall person find it comfortable?'

'It's an individual project,' Alex says. 'A large person could get a print-to-size Sarco. But it depends on which route Philip is going to take. If you have a clinic, we may end up with one-size-fits-all.'

'That's what has to happen in Switzerland,' Philip nods.

Philip is very excited about Switzerland. Exit is going to open a clinic there, the first place in the world where you can be helped to die in a completely non-medical setting. They will be able to provide people directly with the machine, without any 3D printing, because it won't matter if they're assisting in Switzerland. He says he's already found premises and recruited staff. 'Switzerland is the only place we can *give* someone Sarco to use. If you want to use it back home in the UK, well, you'll have to print it.'

'How long did it take to print out?'

They look at each other, smiles twitching.

'Shall we say it?' Alex laughs. 'It took a little while. We were continuously printing for four months.'

'Wow,' I say. 'So it's a peaceful death, at a time of your choosing, so long as you are planning very far in advance.'

'Yeah, this doesn't lend itself to the impetuous user,' says Philip dryly.

They won't tell me how much it cost them to print out, other than it was 'too much,' and funded by 'some big Exit donations.' To be fair to Philip, this isn't something he imagines people rushing to print any time soon. He thinks Sarco will be in widespread use by 2030, when he expects large-scale 3D printing to be commonplace and affordable. But it will still be printed in sections; the frame, the body panels and other components will all need to be assembled, it turns out. And then there's the gas.

'Where do you get the liquid nitrogen from?'

'You buy it,' Philip says, wearily.

'From where?'

'Er, a liquid nitrogen seller,' he scoffs, as if everyone has one on their high street. Perhaps Max Dog will have its own range soon. 'There are plenty of them around and it's not a restricted product in any way,' he adds.

After you print it out, pour in the nitrogen and punch in your code, there are more buttons inside Sarco to make it work: a green 'die' button to initiate the gas, and a red 'stop' button, which you can press if you change your mind. (They can only be pressed from inside – a safety feature intended to prevent Sarco being used to murder someone.) There's also an escape hatch you can push if you feel so inclined, but it doesn't sound like there would be much time to decide to do so.

'You'll lose consciousness in a minute,' Philip explains. 'If you breathe normally, you get into a disorientated state very quickly, have a feeling of euphoria and intoxication, lose consciousness, and then you will be dead in five minutes.'

But Alex says intentionality is built into the design. 'It is surrounded by a perimeter of roughness, it holds you back, tells you, "Think again."' He puts his palm up, like a cop controlling traffic. 'And then there's also softness, so maybe you want to get closer to it, and it's something you are acquainted with, because it looks like a car, but it's a strange car because it's asymmetrical. You can't get in here –' he points to the side without the control panel, the driver's side in the UK – 'because there is no door, so you have to go around. You have to do something yourself in order to go further to the next step that brings you closer to dying by yourself in Sarco. Sarco empowers people to decide. It tells other people that the decision made was the right one, the one the person who ended up inside it wanted.' Sarco's instructions have to be intuitive, for legal reasons. 'If you have to explain to them how to do it, then you are helping. You need the machine to tell you.'

But Philip hasn't just made Sarco so he can get away with helping people die. He's going to use it to make death sexy. 'I like the sense of style, the sense of occasion, the opportunity to redefine death and make it into a ceremony, as opposed to something that you sneak away from and do in private. That doesn't suit everybody, but there are a lot of people that it does. It's a very nice looking device, and something that you can take out so it's looking out over the Alps or the North Sea or the deserts in Australia. The place where *you* want to go.'

'This isn't necessarily about dignity in dying, it's about death being an event?'

'Yeah,' he says, nodding slowly. 'It seems to appeal to a certain group of people. The people who are making contact with us now saying they want to use Sarco see it as something that gives them the chance to mark the event, in the way that sitting in a room and drinking a glass of Nembutal doesn't. This provides a sense of occasion, that they are leaving and travelling. Some people like the idea of saying bye, pulling down the door: "*I'm moving, you're staying.*"' They sound like the sort of people who would love to attend their own funerals.

Sarco also has the allure of the 'euphoria' Philip keeps going on about, of dying high. He says he experienced the intoxication of hypoxia himself, during his air force days, when he lived through a rapid plane depressurization. He had a good time.

'It's horses for courses. I'm not saying everyone is going to want to climb into a Sarco. Some people say, "I don't like the idea; I want to be able to hold the person I love when I die," and this doesn't allow that,' he continues.

'You could print one of these to take two people, like you could print it for tall people,' Alex interjects, helpfully. 'All of that is a possibility.'

'But with two people how do you make sure that both people consent to dying?' I ask.

'It's only a software problem – they both have to pass the test,' Philip says.

'But how can you tell that it's not just one person putting in the codes?'

Philip grits his teeth. There's a ten-second pause. Then they both fall about laughing.

'End of interview!' Alex shouts. 'Cut!'

Surrounded by designers in Venice at sunset, it would be easy to excuse how poorly thought out Sarco is, and treat it as a think piece, a talking point, just like the *XXXXXL Sponge*. But this is not Oron Catts' frog meat. This has been billed as a viable design, funded by people who are desperate to take control of their own deaths, and it is actively being promised to paying

Exit members, who are flooding Philip with enquiries. This is not a joke.

'In ten years' time, do you really expect people to be dying inside Sarcos, all over the world?' I ask Philip.

'I think something like this will be well accepted.'

'It's better than a bag,' Alex adds gently.

'Technology is changing the face of the world, and death is no exception. We're going to see a lot more of people taking control of the final aspects of their life. People are saying, "Enough is enough," in terms of the ability of modern medicine to keep people alive.'

'But then, is the answer a machine to kill them, or a change in our attitude to death?'

'They go hand in hand,' Philip says.

Alex is a relative newcomer to the business of death. 'Have you thought about how you'll feel the first time someone uses your design to kill themselves?'

'Philip will have made the decision to give them access to it, and I trust in Philip to make those decisions,' he replies with a shrug. 'Our responsibility stops with the design.'

Alex suggests I get myself a glass of Prosecco – it's made locally, he says, particularly good here. I head back down to the reception, where the pyramid of fruit has been eaten but the drinks continue to flow. I have a glass on the decking, beside the Grand Canal. While the live band takes a break, Ella Fitzgerald and Louis Armstrong sing 'Cheek to Cheek' on the sound system. 'Heaven. I'm in Heaven.' Everything is balmy, rose tinted, beautiful, unserious, fun.

Except that it's not. It's grotesque. The people who funded Philip's trip and the invention that brought him here aren't thinking about flying joyfully off to the next world in style; they are living with despair, fear, sorrow, pain and panic in this one, and are searching for anyone who can help them out if it. Sarco's launch feels so much like an indulgence, another milestone to flatter Philip's ego, rather than a viable way of helping those people.

Even if the prototype I saw upstairs was in perfect working order and ready to go, it would not be the answer for people who are desperate for a death they completely control. Philip controls this technology, and the access to it. He owns the IP, and if you want it you will have to be accepted into his organization, and pay him.

But then I think about one of the last things Philip said to me upstairs. 'We're planning to make it open source,' he told me. 'We'll make it available to people who have *The Peaceful Pill Handbook*, and that will mean you have to be over a certain age, you'll have to sign something.' He shrugged. 'Look, we know it will bleed. And that doesn't really matter.'

He knows he will never be able to fully control who gets access to the technology he has invented. So long as everyone knows he's the man who created it, he doesn't really care.

CHAPTER FIFTEEN

'The means to an end'

What is it with men and car analogies? RealDolls are the Rolls Royce of adult toys. DS Dolls are the Bugatti Veyron. Clean meat is the automobile that makes the horse and cart of animal meat obsolete. Sarco is the Tesla of death machines.

But Philip wants everyone to know that the true inspiration behind Sarco isn't actually a vehicle at all, but a cult movie released in 1973 starring Charlton Heston.

'I must say that some of my original ideas came from watching the death scene in *Soylent Green*,' he'd told me as he sipped his beer in Venice. 'This futuristic idea that there will be people – and we have them contacting us right now – who say, "I've got to this stage where my life is complete, and I want to do the right thing by the planet."'

The reference was lost on me that day, but in the weeks after the launch I keep hearing *Soylent Green* mentioned whenever Philip talks about Sarco. He gushes about the 'ground-breaking' film in a piece he's written to promote Sarco in the Huffington Post, and in a brief interview with *Vice*, where he repeats that strange idea of a 'death that does what's right for the planet.' So I buy a second-hand DVD to try to work out what he means.

Set in a stinking and violent New York of 2022, where the city's population is forty million and the temperatures are sweltering, *Soylent Green* is the well-worn story of a hard-bitten cop (called Thorn, Heston's character) trying to solve a murder and inadvertently uncovering a global conspiracy along the way. Soylent Green is the name of the lab-engineered superfood humans are forced to eat now that overpopulation and global warming have made conventional agriculture almost impossible. Billed as 'the miracle food of high-energy plankton', it could be any number of the comestibles being cooked up in Silicon Valley today.

The 'death scene' that inspired Philip comes in the final act of the film. Thorn's best friend and flatmate, Sol, who is old enough to remember the good old days, goes to a creepy building where people with benevolent smiles ask him what his favourite colour is ('orange') and his favourite kind of music ('classical'). Then staff in white robes with orange fringing link arms with Sol and guide him to a raised, tomb-like bed – a sarcophagus – where he is propped on a pillow and tucked under some sheets. Bathed in orange light, Sol drinks a cup of something. A button is pressed. Images appear on giant screens around him – orange tulips, orange sunsets, a babbling brook, tropical fish, mountains and a glade carpeted with daffodils – as Beethoven's Symphony No. 6 is piped into the room.

Sol dies with his eyes wide open. The screens and orange lights are switched off. Then the people in the robes wheel his corpse into a chute, which sends his body to the Soylent Green factory, where it is turned into food. Because it turns out that the secret ingredient of Soylent Green is not plankton, but human flesh. 'They're making our food out of people!' Heston shouts in the closing shots of the film. 'Soylent Green is people!'

I blink as the credits roll. Out of all the euthanizing scenarios dreamed up in the canon of science fiction, from *Star Trek* to *Futurama*, this is what has inspired Philip? The calm, controlled death *Soylent Green* depicts is the compliance of an old, depressed and desperate person relieving the burden on an overpopulated planet; death engineered so that humans *can eat other humans*. It is total madness. Can Philip really have seen this cautionary tale and concluded that the death scene is 'doing the right thing by the planet'? Yes, Sol felt no pain, chose when to die, and had his favourite colour shone into his face. But his death was hellish.

When Philip talks about reclining in the Sarco for the sake of the planet, he's describing something eerily similar to the ethical suicide parlours in Kurt Vonnegut's short story, 'Welcome to the Monkey House'. In Vonnegut's fictional world of seventeen billion, the government strategy to address overpopulation included 'the encouragement of ethical suicide, which consisted of going to the nearest Suicide Parlour and asking a Hostess to kill you painlessly while you lay on a Barcalounger.' Perhaps this is what rational suicide is, at its most brutally rational: as soon as you feel you have fulfilled your purpose on earth, the logical thing is to check out as soon as possible and stop taking up precious resources.

We are closer than ever to having to make choices like this. Defying death has become a key objective in Silicon Valley: venture capitalists funding anti-ageing research see a future when death is something that we actively choose when we are tired of living, instead of the scary, unpredictable shadow hanging over us that it is now. Even if escaping death may be beyond us, it's likely that our lifespans, in wealthy countries at least, will stretch out over hitherto unimaginable horizons. Sarco looks like it has been designed not for the terminally ill, but for those who are fit enough to manage to contort themselves into its seat: people who are tired of life and make the choice to die. And because the parameters of illness and disability will no longer be relevant in the decision to give people access to this kind of death, because this will be a death with no gatekeepers, being sure that the choice to die is a rational one, entered into freely, becomes more important than ever.

Which bring us to the mental-capacity assessment required to get the code to enter Sarco, the test that Philip blithely waved away as something that will be carried out by AI as soon as the intransigent medical establishment gives way to the inevitable march of progress. On the face of it, you could easily develop a program that could test whether someone understands what they are about to do when they use Sarco. The Deliverance software already did that part quite effectively: its first question was, 'Are you aware that if you go ahead to the last screen and press the *Yes* button you will be given a lethal dose of medications and die?' and the second was, 'Are you certain you understand that if you proceed and press the *Yes* button on the next screen you will die?' Pretty unambiguous.

But for a person to truly have rational capacity to make a decision, they need to be able to weigh it up and put it in the appropriate context. When doctors evaluate whether someone is fit to choose for themselves they make a value judgement: they look at how the person behaves as much as what they say, not only while taking the test, but in the days and years preceding it. They don't have to agree with the decision their patient makes; they just have to be confident that it was rationally made, based on their answers, behaviour and medical history. It is an art, as much as a science. This value judgement might epitomize the 'doctors know what's best for you' attitude that Philip abhors, but it is the only thing we can rely on for the foreseeable future. In a case where there is any complexity, it's unlikely that computers will be able to get it right, and certainly not by 2030, the time that Philip expects 3D printers to be able to pump out Sarcos quickly and affordably. Getting this right every time really matters, because it's always a life-or-death decision.

Software is not neutral; AI always contains the biases of the people who programmed it, and anything that gets Philip's blessing will be as value-loaded as any doctor's assessment could be. The view that everyone should have the means to have a peaceful death at a time that only they choose is a libertarian position, a political belief, not a fact. With his technology, Philip is able to impose his worldview without any state or doctor getting in the way, and he is imposing it on bereaved families as well as the people dying in his machines. You could say he is as paternalistic as any of the patronizing doctors he despises.

The most revealing insight into how extreme Philip's views are on the right to die comes from his reaction to news of the death of Noa Pothoven. Noa was a Dutch teenager with a history of self-harm, anorexia, depression and post-traumatic stress disorder after being sexually abused at eleven and raped at age fourteen. On 4 June 2019, Daily Mail Online reported that Noa had been legally 'euthanized at home by "end-of-life" clinic' at only seventeen because 'she felt her life was unbearable due to depression'. It was the top story on the site, and went on to make headlines from Australia to India, Italy and the US.

A gleeful press release from Philip popped into my inbox the following day. 'Netherlands Shows Nuance of Euthanasia Debate as Psychiatrically Ill Dutch Teenager Dies', read the headline. 'The global news today that Arnhem teenager, Noa Pothoven, has been helped to die with euthanasia shows the sophistication of the Dutch euthanasia debate as it has developed over the past two decades. Today, I live in a country that is the world leader in open-mindedness when it comes to end of life decision-making for all', Philip gushed. 'There are no hysterics about whether she was sick enough. She was not sick at all. At least not physically. There is little controversy over the fact that she had a mental illness [. . .] *her* opinion of *her* suffering [has] been respected.'

But the story wasn't true. Hours after Philip issued his statement, it emerged that Noa had died at home after refusing food and fluids, and no one assisted her death. Noa had approached a euthanasia clinic without her parents' knowledge in 2017 that refused to help her die. 'They consider that I am too young,' she told the *Gelderlander* newspaper, six months before her death. 'They think I should finish my trauma treatment and that my brain must first be fully grown. That lasts until your twenty-first birthday. It's broken me, because I can't wait that long.'

Amid the flood of international interest, Dutch Health Minister Hugo de Jonge announced an investigation into Noa's death. 'We are in touch with her family, who have told us that there is no question of euthanasia in this case. Questions about her death and the care she has received are understandable, but can only be answered once the facts have been established,' he said.

Philip later wrote a corrective blog post saying he got the story wrong but it didn't matter. 'There is something about the Netherlands that makes the fake news of how Noa died not that relevant [. . .] [T]he fact that her parents allowed her to go through with her wishes, and that the medical profession (in hero role) did not rush in demanding she be saved from herself, says something about this place. The type of respect shown to Noa, if not by not helping her, then at least by not interfering, is a good lesson to those countries who insist on "nanny-stating" the rest of us to death . . . so to speak. Rational suicide is a fundamental human right.'

I believe in the right to die. I believe that future generations will look back in horror when they see how we allow desperate people to suffer – and how people like Lesley, moved by nothing more than love and compassion, are put under immense pressure to break the law to help them – when all they desire is a peaceful and dignified end. But I do not see how there can ever be any 'good lesson' to draw from the starvation of a traumatized, anorexic, self-harming child.

Philip believes anyone should have the right to die painlessly, at a time and place of their choosing, even if they are still in the middle of trauma treatment, as Noa was, even if their brain is still developing, even if there's good reason to think they might one day feel differently. Any psychiatric test that is a barrier to the information and technology he provides is meaningless if Philip thinks people who are profoundly mentally ill are rational enough to choose to die. Sarco's keypad is a fig leaf, the disclaimer that allows Philip to promote his machine while accepting no responsibility for whoever uses it. It doesn't really matter if an AI sophisticated enough to replace psychiatrists is far down the road; Philip wants everyone to have access to his machine anyway, even if there is hope that they might one day want to live.

*

I find Lesley in her new home in rural Norfolk, a cottage surrounded by fields. She's doing bits and pieces of creative writing and is very involved with the local RSPB. Her days of teaching people how to kill themselves are well behind her. Her time with Exit is now little more than a bewildering memory.

'It was looking great,' she tells me in her sun-drenched living room. 'When you go to an Exit meeting it's very obvious that the people there are finding a great relief in being able to talk to people. They can't acknowledge to anyone else that they're thinking of euthanasia at all, so the freedom to be able to speak freely in a safe environment seemed such a great thing.'

She says she had the idea of organizing roadshows, so members around the country could connect with each other, and Exit HQ in Australia seemed keen, but what they really wanted was more members. 'I was told to sign up as many people as we could, encourage people to subscribe to the handbook, sell books and other merchandise, and generally keep the income coming in.' A sad smile crosses her face. 'When I took the job, I didn't think that I would be taking on a sales role.'

Lesley began to question what Exit members in the UK were getting for their money. After the Brayley business Philip had come under scrutiny from the Met Police, which meant Lesley couldn't promise there would ever be any practical workshops from him. 'I was concerned that Exit had actually courted the publicity that had caused this to happen. They were always very pleased when there was anything in the newspapers or the news over here that seemed to make Dr Nitschke into an even more infamous figure. But I was dismayed by the impact that had on what we could then do for members.'

As well as answering the phones to suicidal people, Lesley says she began to field complaints from customers who had ordered equipment through Exit that never arrived, people who had in some cases waited a year or more. She lobbied on their behalf and got them all refunds. But they didn't really want their money back. They desperately wanted someone to deliver on the promise of the peaceful death Philip had sold them. There was nowhere else for them to go.

The main problem was the distribution of Max Dog nitrogen: Exit couldn't find a courier prepared to ship canisters of compressed gas from Australia to the UK affordably. But then a UK nitrogen supplier was found, a company in Margate, which sold canisters to Exit for £43 a pop. Exit then sold them on to British members for £465.

'That did include freight costs,' Lesley adds, apologetically.

'It's an enormous markup,' I say.

'Yes it is. Yes.'

'And people thought they were getting an Exit product, because it was branded as Max Dog nitrogen?'

'They had stickers put on them to say they were Max Dog cylinders, but people knew that they were sourced in the UK, so I don't think that was any kind of deception.' She shifts in her seat. 'It does seem like a huge markup, but Exit does need consistent income, and they'd spent a lot developing the Max Dog range of products. So I was happy, initially, that this was the case.'

'How do you feel about it now?'

Lesley frowns. 'I do accept that they have to cover their costs or they will go under. But I think that the markup was taking advantage of people's need and desperation, in some cases: they knew that individuals wouldn't be able to get hold of these cylinders themselves – because of their age, their infirmity, or for whatever reason it wasn't that simple – and that people would have to buy them through Exit, with a bit of loyalty thrown in to support the cause. They paid a really very high price.'

Even with the new supplier in place, Exit could not find a sustainable way of distributing the cheap nitrogen around the UK. While Lesley was in charge, she says they only managed to ship three canisters. She has no idea if the people who bought them have used them to end their lives.

Lesley and Exit parted ways barely six months after she became the UK coordinator. 'There was such a difference in what I believed the members deserved to get, and what they were actually getting. Philip was very keen that the UK should keep going forward, and we did try to find common ground, but it really wasn't there.' Her contract ended by mutual agreement, she says. 'I'm so disappointed that it hasn't turned out to be what I believed it to be. I genuinely thought that they were doing a great thing for an awful lot of people. Having got to know the ins and outs of the organization more, I can't say I believe that the members are very high on the list of priorities. I think a lot of people have been abandoned, and feel let down.'

In Berkshire, David is feeling better; the NHS has managed to diagnose his mystery digestive problem. 'It was plain sailing from then on. We found the right medication and all is good.'

We sit in his living room, beside his enormous TV, surrounded by ornaments he has collected from his travels overseas. He's a bit anxious; his daughter will be coming home soon and he doesn't want to have to explain to her why there's a journalist on his sofa. But he still very much wants to talk to me – not because he is depressed, this time, but because he is angry.

'Exit has been such a disappointment. The more that I witness, the more I find myself questioning the motives behind it. They are very successful at creating publicity, but given that there's no infrastructure in the UK or supply chain, you have to ask, what is the publicity for?'

David had done all the things Exit members are supposed to do. He bought *The Peaceful Pill Handbook* and ploughed through it. He took out membership so he could attend workshops and chapter meetings. That was the easy part: he just had to give them his credit card details and fill out a form saying how old he was. He says they did nothing to check his age, or the state of his mental health. And David did get the information he was looking for.

When we first spoke, he told me he knew Exit were charging inflated prices, but he didn't mind paying extra, because he believed in Philip. But then he started to have doubts.

'They're dealing with people at their most vulnerable, and people who will do almost anything to achieve their goal,' he says.

'You were feeling pretty low when you found Exit, weren't you?'

He can see where I'm going with this and won't have it. 'This, to my mind, has nothing to do with depression,' he retorts. 'It is my fundamental belief that everyone should have the right to choose when and where they die. I think the tendency of the anti-euthanasia groups to point the finger at depression and make it a reason for not allowing it is wrong. Yes, absolutely, there were times when I got depressed. I was never at a point where the depression took over. I'm not underestimating the power of depression. But being depressed doesn't necessarily drive you to suicide.'

The thing that's really made David angry is the Destiny machine. 'It sounds like the panacea, it sounds amazing. You send off £200, and you get this machine, thank you very much, all your problems are solved. But when you look more closely, the machine seems to depend on a whole collection of ancillaries that you have to have with it to make it work. You need a canister with a mixture of gases in it that currently doesn't exist.' He's talking about the carbon monoxide/nitrogen mix both Destiny and the Mercitron used. 'Even if it were to exist, when you compare it to the nitrogen Exit International are selling, that costs hundreds of pounds. On top of the price of the £200 you've paid for your Destiny machine. According to *The Peaceful Pill Handbook*, it's never been used. It's unproven technology. But it was massively, massively hyped.'

Amid the publicity when Philip unveiled Destiny in Edinburgh, David wanted to find out if he could be one of the first customers to buy it. 'I wrote to Exit International on at least two occasions, enquiring how the whole system worked, what was included, what wasn't included, what you'd have to buy. And unfortunately I got ignored.' He thinks the machine was never anything more than a stunt. 'It was just about raising the profile of Exit International. They want to increase membership. They want people to subscribe to the handbook. That kind of publicity can only help them. Especially since the right to die bill went through the Commons – that was such a mild proposal, and there was such a massive vote against it, so it probably won't be looked at again for quite a few years.'

Philip readily admits that nobody has ever used the Destiny machine to die. He's cited vague 'legal reasons' that mean the project can only remain a prototype. Perhaps Sarco will come to nothing, like Destiny and the CoGen, beyond generating headlines. But I'm not so sure. Philip's plans for Sarco seem so much more concrete. He told me he has premises secured in Switzerland for Exit's new assisted dying clinic where Sarco will be 'the centrepiece', due to open in a few months. Sarco 2.0 is being printed as we speak – this one has a base you can actually pour nitrogen into. And Exit have already been sending out press releases with the name of the first person in line to use Sarco in Switzerland: a forty-one-year-old American woman with MS, called Maia Calloway.

David hasn't renewed his membership. He doesn't need to; he found out what he needed to know and has managed to cobble together his own suicide kit using suppliers he found online that have nothing to do with Exit. I guess that's the flaw in Exit's business model: if it successfully fulfils its members' needs, its membership numbers will necessarily fall away.

David likes talking about his equipment. 'You have to do your research,' he says.

'Everything you've bought is legal, and bought from legal sources?'

'Totally legal.'

'Was it challenging, getting hold of all of it?'

'Yes it was. I had to import some things from abroad. It's a bit like a jigsaw puzzle. You have to put the different elements together to make it work. I have a technical background, and even I have struggled with certain aspects of putting this together. I think the majority of Exit members have no understanding of the mechanics, and essentially want to buy an off-the-shelf kit which will deliver what they want with an instruction manual, a flat-pack instruction manual – plug A into B, do C, there's your result.'

He leads me up the stairs to his bedroom, on the top floor. There is a wardrobe near the door. He stoops to pull out a tangle of tubes, canisters and regulators from some low hiding place. It's all a bit rushed; he really doesn't want his daughter to find us here now, but he's proud of what he's achieved, and he wants me to see it.

'That's everything you need to take your own life?'

'Yes, right there in my cupboard.'

I try to imagine being able to sleep soundly when I'm a metre away from the device that will one day kill me.

'Doesn't it make you feel uncomfortable, knowing that you have this in your bedroom?'

'No,' he says, firmly and deliberately. 'It's my comfort and my insurance policy. It gives me peace of mind. Many, many people have a fear of becoming old, becoming sick, becoming incapable, becoming a burden to other people. Lots and lots and lots of people don't want that to happen. If you can provide yourself with the means to an end – literally! – that will, at the time of your choosing, prevent you becoming a burden to someone else, it takes away the fear of the future.'

David does not need a death machine. He needs to live in a world where ageing, disease and death are no longer terrifying; a world where we learn to live alongside our mortality, and are prepared to confront illness and death as a natural part of life. For that, we need proper investment in research into dementia, motor neurone disease and the other conditions that strike so much fear into our hearts; we need better funding of palliative care and social care, so that no one can ever consider themselves a 'burden'. Because the people who desire to have control over their own deaths often really want dignity and reassurance, not death itself.

And, more than anything, we need the right to die to be enshrined in law. We need to find a way of legalizing assisted dying without endangering vulnerable people who want to live. That will take more intellectual effort than designing a death machine, and it won't make anyone rich or famous, but until we get it right, desperate people will be left open to exploitation.

*

Maia Calloway is not hard to find. She's left her email address in the comments section of a blog post she's written about the right to die, and when I send her a message she replies within minutes. 'I'd be more than happy to speak to you and contribute in any way,' she writes. 'I'm fascinated with the Sarco and what it represents.' We make a plan to Skype the following day.

Exit has been sending out regular press releases mentioning Maia. One arrived on the day of the Venice launch, and I read it on the water bus from the airport. It included a picture of

Maia smiling on a bench, with a delicate face and ice-blue eyes, a striped shawl draped across her slender shoulders. It mentioned that Maia had already travelled to Switzerland for assisted dying before, but had decided to return to the US.

Now, almost a year and a half later, Maia thinks her time is near, it read, in breathless bold and italics. *And she wants to use Sarco.*

Philip brought Maia up when I spoke to him later that evening.

'I saw the press release,' I said. 'She went to Switzerland and changed her mind?'

'She didn't change her mind so much; she realized, because of her multiple sclerosis, it was a slower process, and she thought, I'll go back to America. But she's coming back again. The only thing is whether her timing suits our timing. If the machine is available, she says she likes the concept.'

'She'll be the first person to use it?'

'If the timing's right,' Philip said, holding up his crossed fingers in almost gruesome anticipation.

When the time comes for our Skype chat, Maia emails to say I'll need to call her phone; she can't work out how to launch Skype without her carer there.

'I'm so sorry,' she says when she picks up, her voice quiet but steady. 'In future, I'll have all this worked out. I'm just having some cognitive problems with the MS.' She has just turned forty-one, she says, yet she feels 'more childlike because of the progression of the illness. It's like I'm getting younger. I want the comforts of a child: always needing hugs and cuddles, needing to have my food prepared for me, needing to be tucked into bed.' Maia lives with her best friend in Taos, a small town in New Mexico, in the southernmost range of the Rocky Mountains. Her mother and sister died during the same period when her MS worsened. 'There wasn't really anyone to take care of me, other than hired help for a few hours a day. So my friend looks after me. He's like a big brother.'

Her openness and her soft voice do make Maia sound child-like. Only minutes into our conversation, a kind of maternal instinct kicks in, and I feel a pang of horror. What is Maia doing in Philip's world? But when I ask her about what her condition is like on a day-to-day basis, it's clear that I'm talking to an intelligent, rational adult. She has the articulate vocabulary of the fully grown, college-educated woman that she is.

'It's just a continuous descent. It's insidious, like a narrow corridor just getting more and more narrow. You don't have dementia like you would with Alzheimer's, but you have severe cognitive impairment, so memory, attention, executive function, being able to learn new tasks – all of those things are severely degraded. Then, with the spinal cord lesions, your arms and legs and torso stop functioning.' Paralysis is inevitable, she says. 'At the point of full paralysis, it's a lot like motor neurone disease, but it goes on *longer*. In a year or two I could be fully paralysed but even at that point I will not be judged terminal or hospice eligible, so the last stretch could be a few years of being completely bed bound, having absolutely no control of my bodily functions, having difficulty communicating.' She is already losing control of her neck, and has breathing problems. 'I didn't even really want to see it to *this* level. I don't want to go too much further.'

When Maia was well, she was a fiercely driven woman. She worked in film production and lived for her career. 'If you asked that former self, "Would you want to live the way I am living now?" I would have said, "Absolutely no." But actually being able to do it is a lot harder than you might think. That survival instinct kicks in.'

'What do you mean by "actually being able to do it"?'

'I mean either doing something on your own, like Philip's *Peaceful Pill Handbook*, or actually going over there and taking the medicine. I have already gone to Switzerland and got the green light, and I came back because I wasn't ready.'

Maia's account of her visit to Switzerland differs from Philip's. It wasn't that she realized the progression of her MS was slower than she had thought; it was more that she couldn't go through with it. She arrived in Zurich on her own. She was evaluated by doctors from the Lifecircle euthanasia clinic, who set her up with a carer for a few days. She saw some of the local sites, visited a monastery. And then she started to feel guilty.

'I think it was really about shame, the way my culture thinks about the shame of suicide. There's so much MS in America, and there's an unspoken agreement in our society that if you have progressive MS, sorry, you've just got to learn to deal with it and keep that fighting attitude. You are kind of a poor sport if you can't see it to the bitter end – you're less brave, less courageous.' And then she began to think about her father. 'There was a whole thing of, You can't let your dad lose another daughter – that's forbidden. You don't go before your elders.'

However it happens, suicide is never an entirely solitary, individual act. There are always other people involved: those who assist you, those who happen to be near you, those who find you, those who love you that you are leaving behind.

'Did your dad know you were making that trip to Switzerland?'

'No. He found out through one of his busybody friends. He got very angry. He felt betrayed. And I felt, Oh God, my dad's mad at me; I'm in trouble. I swiftly got back on the plane and came back to my friend who takes care of me. The agreement was that we were going to press on a little longer and do this the right way, giving everybody in the family the information that they deserved, and then hopefully having somebody escort me back when I was absolutely ready. But the irony is, having come back, *nothing* got accomplished. They do not want to accept this. They don't want to talk about it. They don't want to take me to the plane. They certainly don't want to go over there. The sad part of my story is that having come back to "do it the right way", their response is still the same.'

It was in Switzerland that Maia met Philip for the first time. 'He's a personal hero of mine,' she enthuses. They had exchanged emails before, and when she found out they were in the country at the same time she asked if he would meet her. 'I went up to Grindelwald with my caregiver and I met Philip and Fiona and their little dog, Henny. It was wonderful. We had a pizza, we talked about a lot of stuff. Then he pulled out on his iPhone pictures of this device, and said, "This is what I've been working on."'

Philip never misses an opportunity. I can picture him now, around the table with his wife, his dog, his new disabled friend and her carer, a slice of pizza in one hand and his iPhone in the other, rolling out the Sarco concept pictures and *Soylent Green* anecdotes. Maia was impressed. 'I thought, Wow, this is absolutely fabulous.' But it didn't look like it was going to be ready any time soon, so she went back to the US and didn't think any more of it.

They stayed in touch. 'I said, "If I can help you at all, Philip, being an American who is denied the right to die, let me promote your cause."' And that's what happened. 'Eventually he said, "Would you like to try the Sarco?" And I said, "Well, I will keep it open, and I will tell the media that I am very interested in it because of the ways the laws exclude me."'

Maia is choosing her words carefully here, because while she's certainly interested in Sarco, she has no plans to die in it.

'I have very decreased respiratory function and a little bit of – what's the thing where you are afraid of small spaces?'

'Claustrophobia.'

'Yeah, I have a little bit of that. I think the Sarco is fantastic. It's beautiful. It's elegant. What it symbolizes is so wonderful for our world. But for me, with my particular illness and anxiety, I don't know that it's the right fit. But I am still fascinated with it, and I think it's what the future is going to be.'

But then, before I can even ask, Maia brings up a host of reasons to worry about Sarco.

'When you see in *Newsweek* that it's the Tesla of death machines, we have to be careful that we don't get so wrapped up in the elegance and the chicness of it that we forget that we are talking about life or death, and this is a process that you have to be very rational about.' Sarco makes death glamorous, euphoric and therefore alluring, but suicide is contagious enough anyway, particularly among the young, particularly when it gets international coverage. In the month after Marilyn Monroe died there was a 12 per cent rise in suicides in the US, and Robin Williams' death was linked to a 10 per cent rise in suicide rates in the five months after he killed himself. Suicide doesn't need a new machine to add to its appeal.

'I'm also a little concerned that it could somehow malfunction with any random printing,' Maia continues. 'You would never know what abnormality could occur.' I hadn't thought about that, even though Alex had readily admitted to me in Venice that printing had been a nightmare because 'the machines tend to fuck up'. A defective machine would be devastating for anyone who has psyched themselves up to use it. Maia has been talking about Sarco with NuTech co-founder Derek Humphry.

'Derek said to me, "Things like this have been tried in the past and have had problems. My advice is that, if you're going to do it and be the first person, somebody ought to be standing by with an injection." And I thought, Oh *shit*.'

The first time anyone climbs into Sarco and presses the button it will be an event. Philip is already drumming up press interest for it. But Maia isn't thinking of her own death as a performance; she isn't an audience member at Philip's Edinburgh Fringe gig having a go on his new death machine for a laugh. She needs to know whatever is being used will definitely end her life. 'I'd have to be absolutely totally sure.'

There are no certainties in Maia's life. She is in limbo, neither unwell enough to die nor well enough to live. But it's the way the world around her responds to her inability to fit into neat categories, her inbetweenness, that makes her existence so unbearable.

'For the degenerative, incurable illnesses, you don't have that kind of compassion that the terminal hospice people have, and you're obviously not healthy and out in the world competing. You are cut off. America is not a society that favours the physically imperfect at all. It's very cut-throat. And certainly the media world, where I came from. When you are scared, imperfect and impaired, it's not a society that embraces you.'

'But isn't the answer to change those attitudes in society, rather than develop a technology to kill you?'

'Yeah. Right! I think we have to work on all fronts.'

Philip said the same kind of thing to me in Venice. But just like the effect IVF had on research into the causes of infertility, the easy answer Sarco provides might make it less likely that we investigate what makes a person want to end their life. And while death remains taboo and assisted dying continues to be an option only open to a select few, there will always be a market for DIY death. Like backstreet abortions, the drive will still be there, regardless of whether there is the technology or legal framework to ensure it can be done in a safe and dignified way.

'To die in my bed, with my Cheshire cat who I love so much, and have a last meal – that's ideally how I'd go,' Maia says. 'But my family dynamic is not healthy. Like so many American families, we are terrified of illness and death. Given my domestic situation it's probably better to be in the peaceful dying apartment by the lake in Zurich or Basel, because it's a space that is very safe and guaranteed, where it is culturally accepted and there is no shame.'

Of all the people I've met who want to be in control of the end of their lives, Maia is the closest to the end. She expects to end her life at the Lifecircle clinic sometime within the next few months. Death is not some insurance policy waiting in her cupboard, a vague concept she is yet to confront – she is staring it in the face.

'Is there such a thing as a perfect death?' I ask her. 'Can it ever exist?'

Maia pauses for a moment.

'Aesthetically it *is* the Sarco. You have an elegant device that actually makes you high and elated before you take off, right? It's in some beautiful setting, because you can take it to your favourite place. That, aesthetically, is the perfect death,' she replies eventually. 'But what is really, profoundly, the perfect death is that you have made amends with everybody, and you are at peace with what occurred in your own life and your own mortality. You have cut ties with those attachments to your personal belongings, your resentments, your addictions, your anger. That is the perfect death for me – understanding and having gone through those steps of acceptance. The Sarco is beautiful, but if you don't have those things in place then you can still be a tormented soul inside it.'

'The perfect death is a state of mind, and not a means of dying?'

'Yes,' she says wistfully. 'Yes, yes, yes.'

Epilogue

As I write this, Harmony isn't on the market yet. Sidore and the rest of Davecat's dolls remain the centre of his world, undisturbed by the artificially intelligent mistress that might one day steal his heart. The JUST chicken nugget still hasn't been released in a high-end restaurant in a country with a relaxed attitude to food regulation. CHOP are expecting the FDA to rule on whether they can begin putting human babies into their biobag sometime in 2020, and they are hopeful that it will be in widespread use by the end of the decade. Wes and Michael have had a baby boy called Duke. smithe8 has deleted his Reddit account and disappeared from the manosphere. Sarco 2.0 is being spewed out in layers of semi-biodegradable plastic by the printer in Haarlem. Maia Calloway won't be the first to use it, but Philip says at least a hundred people are in line behind her to die within its lacquered shell.

In other words, none of the innovations I've encountered really exists yet. Harmony, JUST meat, the biobag and Sarco may all be suffused with hype, but the solutions they promise to provide are too alluring for them not ever to exist, the commercial imperative too great. They will go on the market one day, even if it's not as soon as Matt, Josh, the CHOP team and Philip might promise.

While their products remain in their workshops, their

competitors are making strides. DS are taking £300 deposits on their first-generation heads. Cloud Climax have started stocking Emma, a £3,000 robotic head from another Chinese company, AI-Tech; she's billed as 'a secretary without temper', who always calls her owner 'Master'. Emma is little more than a winking, blinking mannequin who can read out your calendar alerts, but AI-Tech promises that 'the more you talk, the more she learns'.

A new artificial womb prototype was unveiled at Dutch Design Week 2019 that does away with lambs altogether. Eindhoven University of Technology's take on ectogenesis hangs from the ceiling, like an enormous crimson beach ball, and comes complete with a reassuring artificial maternal heartbeat. The Dutch team will test it using 3D-printed fake babies fitted out with a vast array of sensors, and plan to move on to live human foetuses as soon as possible. In October 2019, the project won €2.9 million in EU funding. Professor Guid Oei, who is leading it, has hailed his invention as a 'gamechanger'.

Clean meat start-ups are springing up across the globe, growing exponentially like starter cells in FBS. The FDA in the US and the British government still can't decide whether clean meat can be called meat, and the clean meat industry is quietly dropping the 'clean' label; it isn't catching on, and it's making the meat industry antsy at a time when everyone wants to keep it on side and investing heavily. (Even Bruce is changing his mind: in September 2019 he announced that the GFI was going to 'embrace new language' and start calling it 'cultivated meat' instead.) But plant-based burgers are taking the world by storm. When Beyond Meat shares hit the stock market it was the best-performing initial public offering of 2019, up 600 per cent in its first month. The Impossible Whopper is now on the menu of Burger King branches across the US, and Impossible is trying to work out how to meet the demand. Animal-free meat is taking off, even if disembodied flesh hasn't quite worked out what it is yet.

Before birth, food, sex and death are changed forever, significant hurdles will need to be crossed. First there is the yuck factor, the ick factor, the uncanny valley, the disgust human beings feel when something as intimate as how they have sex, what they eat, how they are born and how they die is challenged by a radically new means of production. The entrepreneurs are finding ways around it, with clever language, emotive arguments and sleek design. The shock of the new is nothing new. And if babies conceived in test tubes can become unremarkable, so can robot wives and babies in bags.

Then there's the question of who will get to use these technologies. They will be elite products, at least at first. For all Philip's bluster about the universal human right to rational suicide, the death Sarco offers is a luxury for the privileged, and as much as Josh is working towards a just world 'guided by reason, justice and fairness', I can't imagine the people he met in Liberia tucking into one of his Wagyu beef patties any time soon. Ectogenesis will only bring equality in reproduction for women rich enough currently to afford a social surrogate, and foetal rescue will only be possible in countries wealthy enough to have it in their arsenal of social care options. Even cut-price Chinese sex robots will cost a significant chunk of disposable income. The men determined to go their own way will need a lot of cash to be truly free of women.

Men dominate the tech industry, and their inventions reflect their egos and desires. But women will be disproportionately affected by the technologies I've encountered, and not just sex robots and artificial wombs. Most of those who died using Kervorkian's machines were female, and wherever assisted dying is

legal women choose it more often than men, even though suicide is generally much more of a male phenomenon. Women
are more likely to outlive their partners, and are more used to
doing the caring than being cared for. It's possible that the fear
of being a burden is felt even more acutely by women. And as
Mark Post told me, 'Meat has always been associated with
power.' Meat is about 'eating like a man'. In every part of the
world, men eat more meat than women. Meat is masculine, and
so is the rampant overconsumption that's causing so much
harm. If the solution is lab grown meat, all of us will become
dependent on ever more specialized technology where we were
once self-sufficient, but women disproportionately so, compared
to their desire for meat in the first place. These innovations tell
us a lot about male appetites for food and sex, and the male
desire to control birth and death.

But both women and men fear disorder and powerlessness.
Human beings want control over our environment, over our
food, our bodies and each other. Sex robots are substitute partners, without the autonomy that makes human relationships so
precarious. Clean meat is a substitute for animals, without the
shit, disease and pollution that could lead to our extinction.
Artificial wombs are substitute pregnant women, without their
fallible bodies and potential for unmotherly behaviour. Death
machines are a substitute for an unpredictable, undignified
death. They are proxies that distance us from our nature, from
the world around us and from each other.

If we agree to outsource food, sex, birth and death to
machines in order to have the illusion of control, we risk losing
hold of our empathy, our imperfections, our agency, the contin-
gency of our existence. Technology dehumanizes us. Even if it
really is being developed with the noblest intentions – *Saving
the planet! Saving tiny babies! Giving companionship to the*

lonely! Setting the sick free! – we have no idea whose hands these inventions will fall into, what they will use it for and where it will ultimately take us.

The 'problems' the innovations in this book are supposed to solve were caused by technology in the first place. Industrial agriculture has made animal meat unsustainable; the pill has given women the independence that is so inconvenient for men who want a partner who exists purely for their pleasure; medical interventions have made gestation inside the female body seem ever more risky; better medicines have made ageing, disease and death seem terrifying. Every time we rely on technological solutions, we risk becoming dependent on different orders of magnitude of complexity to carry out what has always come to us naturally. We disempower ourselves, and lose part of ourselves.

None of these inventions are actually solutions: they are circumventions. Instead of looking at why some of us have the desire to have partners with no autonomy, to have babies without being pregnant, to eat large amounts of meat even though it damages the planet and our bodies, or to be in total control of our own deaths, the people I've met are selling us a way to ignore natural human anxieties. Instead of setting us free, they help us live with the conditions that are trapping us in the first place. They depoliticize them, obscure and bypass them. They're giving us reasons not to know ourselves better.

What does this mean for all of us? It can mean whatever we want it to. At its most dystopian, it means that women could become obsolete, that empathy will become hard work, that multinational companies could have total control of the meat industry, that vulnerable people will be able to download their own death with no oversight whatsoever. But that would be to take a fatalistic view of human nature, one I don't buy.

We can use the time we still have, before these inventions go on the market, to examine why we think we need them in the first place. Then we can make the changes and sacrifices necessary to solve fundamental human problems, instead of turning to technology to paper over them. We *will* have to make sacrifices – we can't have our steak and eat it, we can't have everything we want without any consequences, no matter what scientists and entrepreneurs may say. These inventions are going to change us, if we are not prepared to change our behaviour.

Progress is the courage to choose a different mindset. That has to come before technological innovation, not because of it. And in some parts of the world we are already making the changes we need to move forward without these inventions. Every year, in wealthy countries at least, more citizens are being given the right to die in a safe and dignified way. Mothers are getting better maternity care and protection for their jobs. More people are becoming vegan or actively choosing to eat less meat, and fewer parents are bringing their kids up as meat eaters. The incels and MGTOWs of the men's rights movement are an eye-catching but tiny minority; most men want their partners, sisters and daughters to be respected, protected and equal.

The people I've met in these pages know this, but they also know that social change is hard work, and there is money to be made in offering an easier fix. It's up to us whether we choose to buy it.

If only everyone had bothered to read until the closing thoughts of that totemic 'Fifty Years Hence' essay Churchill wrote in 1931: 'Projects undreamed of by past generations will absorb our immediate descendants; forces terrific and devastating will be in their hands; comforts, activities, amenities, pleasures will crowd upon them, but their hearts will ache, their lives will be barren, if they have not a vision above material things.'

I've been trying to find out what these projects undreamed of will mean for our immediate descendants. One person's dystopia is another's bright future. But the words that have lingered most for me didn't come from Matt McMullen, or Mark Post, or Anna Smajdor, or Philip Nitschke. They were said by perhaps the most unassuming person I met.

As I was packing up my notebook on that cold day at the Open University in Milton Keynes, Matthew Cole, the vegan sociologist, drained the last of his coffee.

'Coming up with technical fixes rather than ethical reform, revolution, rebellion . . . Every time that technology tries to stand in for ethics, we do ourselves a disservice,' he said. 'We deny ourselves the opportunity for growth.'

It's not possible to live a selfish life with a completely clear conscience, but living alongside imperfection, compromise, sacrifice and doubt is as fundamental a part of the human experience as birth, food, sex and death. We can choose whether to accept the messiness of our existence, or we can continue to try to use technology to cancel it out, like the earplugs in my Las Vegas hotel room. We don't need sex robots and vegan meat. The freedom and power they promise us are already in our hands. We already have the answers. Putting them into action will take so much more than just opening a bag, closing a door, or flicking a switch.

Acknowledgements

I am incredibly grateful for the generosity of the people who let me interview them for this book. Many had no idea that talking to me would take up so much of their time. Thank you, and sorry I was in your hair for so long.

I'd also like to thank:

My agents, Sophieclaire Armitage and Zoe Ross, for their support and ideas, and for immediately getting what I wanted to do.

My editor, Kris Doyle, for the enthusiasm, for the clarity of vision and for the title. James Annal, for the wonderful cover design. My publicist, Anna Pallai, for her determination in such extraordinary times.

The people I rang up and disturbed far too often from their much more important work: Julie Kleeman, Rick Adams, Sarah Eisen and Saul Margo. Thank you for your expertise.

My colleagues at the *Guardian*: much of this book was made possible because of research I did for pieces that first appeared there. Enormous thanks must go to Tom Silverstone, who did so much to make my work into sex robots come alive, and to Mike Tait and Mustafa Khalali, who commissioned the film Tom and I made together. Thanks also to Clare Longrigg, Jonathan Shanin, David Wolf, Charlotte Northedge, Ruth Lewy and

Melissa Denes, whose razor-sharp editing of those pieces taught me how to write.

The people who looked at very early drafts: Rick Adams, Ed Reed and Elizabeth Day. Stig Abell, who told me I should be writing a book in the first place.

Laura Solon and Dan Pursey in Los Angeles, and Olivia Solon and Stu Wood in San Francisco, who fed me, gave me coffee and let me sleep in their spare rooms.

My parents, David and Manou, and my sisters, Susanna, Nicole and Julie. Where do I begin? I am so lucky to have you all.

Anna Kehayova, who held my life together while I was writing this. I can't thank you enough.

My children, who by and large stayed out of my bedroom when I was in there writing this.

Scot, my partner in everything, and the smartest person I know.

But the greatest thanks must go to Corrie Bramley, to whom I owe so much. Without her, every page of this book would be blank.

Notes

Chapter One

p. 24 **worth over $30 billion**   This is according to the entrepreneur and investor Tristan Pollock, when he was at 500 Startups. *See* Andrew Yaroshenko, 'What is #SEXTECH and how is the industry worth $30.6 billion developing?', 4 June2016,https://sexevangelist.me/what-is-sextech-and-how-is-the-industry-worth-30-6-billion-developing-d5f0a61e31d6

p. 24 **a 2017 YouGov poll**   Yael Bame, '1 in 4 men would consider having sex with a robot', 2 October 2017, https://today.yougov.com/topics/lifestyle/articles-reports/2017/10/02/1-4-men-would-consider-having-sex-robot

p. 24 **A 2016 study**   Jessica Szczuka and Nicole Krämer, 'Influences on the Intention to Buy a Sex Robot', 18 April 2017, https://www.researchgate.net/publication/316176303_Influences_on_the_Intention_to_Buy_a_Sex_Robot

Chapter Two

p. 34 **Pygmalion**   This idea has become folklore among people who study sex robots. See David Levy, *Love & Sex with Robots* (HarperCollins, 2007) and Kate Devlin, *Turned On* (Bloomsbury Sigma, 2018).

p. 34   **Laodamia**   Kate Devlin's *Turned On* (Bloomsbury Sigma, 2018) looks at the history and pre-history of all this in great detail, and it's an entertaining read.

p. 39   **Fox News**   'ROXXXY, the World's First Life-Size Robot Girlfriend', Fox News, 11 January 2010, http://www.foxnews. com/tech/2010/01/11/worlds-life-size-robot-girlfriend.html

p. 39   ***Daily Telegraph***   Andrew Hough, 'Foxy "Roxxxy": world's first "sex robot" can talk about football', *Telegraph*, 11 January 2010, https://www.telegraph.co.uk/news/newstopics/ howaboutthat/6963383/Foxy-Roxxxy-worlds-first-sex-robot-can-talk-about-football.html

p. 39   ***Spectrum***   Susan Karlin, 'Red-Hot Robots', *IEEE Spectrum*, 15 June 2010, https://spectrum.ieee.org/robotics/humanoids/ redhot-robots

p. 39   **ABC News**   Ki Mae Heussner, 'High-Tech Sex? Porn Flirts With the Cutting Edge', ABC News, 8 January 2010, https:// abcnews.go.com/Technology/CES/high-tech-sex-porn-flirts-cutting-edge/story?id=9511040

p. 39   **CNN**   Brandon Griggs, 'Inventor unveils $7,000 talking sex robot', CNN, 1 February 2010, http://edition.cnn.com/2010/ TECH/02/01/sex.robot/index.html

p. 41   ***New York Times***   Laura Bates, 'The Trouble With Sex Robots', *New York Times*, 17 July 2017, https://www.nytimes. com/2017/07/17/opinion/sex-robots-consent.html

p. 41   ***The Times***   Kate Parker, 'A sinister development in sexbots and a strong case for criminalisation', *The Times*, 21 September 2017, https://www.thetimes.co.uk/article/a-sinister-development-in-sexbots-and-a-strong-case-for-criminalisation-qxxxjkmsl

Chapter Three

p. 57   *Fortune*   Jonathan Vanian, 'The Multi-Billion Dollar Robotics Market Is About to Boom', *Fortune*, 24 February 2016, https://fortune.com/2016/02/24/robotics-market-multi-billion-boom/

p. 63   *New York Times*   Ross Douthat, 'The Redistribution of Sex', *New York Times*, 2 May 2018, https://www.nytimes.com/2018/05/02/opinion/incels-sex-robots-redistribution.html

p. 63   *Spectator*   Toby Young, 'Here's what every incel needs: a sex robot', *Spectator*, 5 May 2018, https://www.spectator.co.uk/2018/05/heres-what-every-incel-needs-a-sex-robot/

p. 64   **They argue that giving men**   Roc Morin, 'Can child dolls keep pedophiles from offending', *Atlantic*, 11 January 2016, https://www.theatlantic.com/health/archive/2016/01/can-child-dolls-keep-pedophiles-from-offending/423324/

p. 65   **The Samantha Project**   Sergio Santos and Javier Vazquez, 'The Samantha Project: A Modular Architecture for Modeling Transitions in Human Emotions', *International Robotics & Automation Journal*, Volume 3, Issue 2, 2017, pp. 275–80.

p. 67   **BBC crew**   *Sex Robots and Us*, BBC Three.

Chapter Four

p. 76   **'consensual nonmonogamy'**   Kate Devlin, 'I have other men. He has other women. We're both happy', *The Times*, 10 June 2017, https://www.thetimes.co.uk/article/i-have-other-men-he-has-other-women-were-both-happy-29wkdjd99

p. 76 **'It can go somewhere else** Kate Devlin goes into a lot more detail about this in her book, *Turned On* (Bloomsbury Sigma, 2018), which explores the past, present and future of sex tech from an academic perspective. Well worth a read.

p. 79 **'monoheteronormative'** *See* Kate Devlin, *Turned On* (Bloomsbury Sigma, 2018).

Chapter Five

p. 93 **becoming more carnivorous** OECD, Meat consumption (indicator), 2018, https://doi.org/10.1787/fa290fd0-en (Accessed on 21 November 2018).

p. 93 **The US alone** National Cattlemen's Beef Association, Industry Statistics, http://www.beefusa.org/beefindustry statistics.aspx

p. 93 **stretch to the moon and back** I've done some creative maths here, but I think it's accurate. Twenty-six billion pounds of beef would make 104 billion quarter pounders, each around two thirds of an inch thick, so, when stacked on top of each other, they would reach 69.33 billion inches. It's 15.13 billion inches to the moon, so they could reach the moon and back twice, with enough left over to go around the circumference of the earth five and a half times.

p. 93 **70 billion animals are killed** Compassion in World Farming, 'Strategic Plan 2013–2017', https://www.ciwf.org.uk/media/3640540/ciwf_strategic_plan_20132017.pdf

p. 93 **The global livestock industry** Food and Agricultural Organization of the United Nations, 'Major cuts of greenhouse gas emissions from livestock within reach: Key facts and findings', 26 September 2013, http://www.fao.org/news/story/en/item/197623/icode/

p. 93 **The world's three biggest meat companies** GRAIN, IATP and Heinrich Böll Foundation, 'Big meat and dairy's supersized climate footprint', 7 November 2017, https://www.grain.org/article/entries/5825-big-meat-and-dairy-s-supersized-climate-footprint

p. 94 **For every 100 grams of beef** J. Poore and T. Nemecek, 'Reducing food's environmental impacts through producers and consumers', 22 February 2019, https://josephpoore.com/Science%20360%206392%20987%20-%20Accepted%20Manuscript.pdf

p. 94 **more than 50 per cent** R. Goodland and J. Anhang, 'Livestock and Climate Change', Worldwatch Institute, November 2009, https://www.researchgate.net/publication/285678846_Livestock_and_climate_change

p. 94 **52 per cent** Chen Na, 'Maps Reveal Extent of China's Antibiotics Pollution', Chinese Academy of Sciences, 15 July 2015, http://english.cas.cn/newsroom/news/201507/t20150715_150362.shtml

p. 94 **70 per cent** '2016 Summary Report on Antimicrobials Sold or Distributed for Use in Food-Producing Animals', US Food and Drug Administration, Center for Veterinary Medicine, December 2017, https://www.fda.gov/downloads/forindustry/userfees/animaldruguserfeeactadufa/ucm588085.pdf

p. 94 **China and the USA combined** UN Food and Agriculture Organisation, 2018, https://ourworldindata.org/grapher/meat-production-tonnes?tab=chart&country=MAC+USA+GBR+CHN+Europe

p. 94 **Pneumonia and tuberculosis** 'Antimicrobial resistance', World Health Organization, 15 February 2018, http://www.who.int/news-room/fact-sheets/detail/antimicrobial-resistance

p. 95 **If nothing changes** Jim O'Neill (chair), 'Tackling Drug-Resistant Infections Globally: Final Report and Recommendations', Review on Antimicrobial Resistance, May 2016, https://amr-review.org/sites/default/files/160525_Final paper_with cover.pdf

p. 95 **The most efficient meat** A. Shepon, G. Eshel, E. Noor and R. Milo, 'Energy and protein feed-to-food conversion efficiencies in the US and potential food security gains from dietary changes', *Environmental Research Letters*, 11, 2016, 105002, http://iopscience.iop.org/article/10.1088/1748-9326/11/10/105002/pdf

p. 95 **43,000 litres** D. Pimentel, B. Berger, D. Filiberto, M. Newton, B. Wolfe, E. Karabinakis, S. Clark, E. Poon, E. Abbett and S. Nandagopal, 'Water Resources: Agricultural and Environmental Issues', *BioScience*, Volume 54, Issue 10, October 2004, pp. 909–18, https://academic.oup.com/bioscience/article/54/10/909/230205

p. 95 **a forty-eight-hour shower** This is using a figure of fifteen litres per minute, which seems to be pretty average for showers.

p. 95 **it takes 112 litres** M. M. Mekonnen and A. Y. Hoekstra, 'The Green, Blue and Grey Water Footprint of Farm Animals and Animal Products', *Value of Water Research Report Series* No. 48, UNESCO-IHE Institute for Water Education, December 2010, https://waterfootprint.org/media/downloads/Report-48-WaterFootprint-AnimalProducts-Vol1_1.pdf

p. 96 **Eutrophication** M. Selman, S. Greenhalgh, R. Diaz and Z. Sugg, 'Eutrophication and Hypoxia in Coastal Areas: A Global Assessment of the State of Knowledge', World Resources Institute, *WRI Policy Note*, No. 1, March 2008, https://www.researchgate.net/profile/Suzie_Greenhalgh/

publication/285775211_Eutrophication_and_hypoxia_in_
coastal_areas_a_global_assessment_of_the_state_of_
knowledge/links/5679c00e08ae361c2f67f4d8/
Eutrophication-and-hypoxia-in-coastal-areas-a-global-
assessment-of-the-state-of-knowledge.pdf

p. 96 **Almost 80 per cent** Food and Agriculture Organization of
the United Nations, 'Animal production', http://www.fao.org/
animal-production/en/

p. 96 **Up to 80 per cent** H. Ritchie and M. Roser, 'CO$_2$ and
Greenhouse Gas Emissions', Our World in Data, December
2019, https://ourworldindata.org/co2-and-other-
greenhouse-gas-emissions

p. 96 **Researchers at Oxford University** J. Poore and
T. Nemecek, 'Reducing food's environmental impacts
through producers and consumers', 22 February 2019, https://
josephpoore.com/Science%20360%206392%20987%20-%20
Accepted%20Manuscript.pdf

p. 101 **The meat and poultry industry** 'New Economic Impact
Study Shows U.S. Meat and Poultry Industry Represents
$1.02 Trillion in Total Economic Output', North American
Meat Institute, 14 June 2016, https://www.meatinstitute.org/
index.php?ht=display/ReleaseDetails/i/122621/pid/287

p. 109 **the growth of veganism** 'Statistics: Veganism in the UK',
Vegan Society, https://www.vegansociety.com/news/media/
statistics

Chapter Six

p. 112 **$1.1 billion** This figure comes from Josh Tetrick himself,
and, as you will soon see, everything he says needs to be
taken with a pinch of salt.

p. 114 'the company used shoddy science Biz Carson, 'Sex, lies, and eggless mayonnaise: Something is rotten at food startup Hampton Creek, former employees say', Business Insider, 5 August 2015, http://uk.businessinsider.com/hampton-creek-ceo-complaints-2015-7?r=US&IR=T

p. 114 **Bloomberg** Olivia Zaleski, 'Hampton Creek Ran Undercover Project to Buy Up Its Own Vegan Mayo', Bloomberg, 4 August 2016, https://www.bloomberg.com/news/articles/2016-08-04/food-startup-ran-undercover-project-to-buy-up-its-own-products

p. 118 **plunging a needle** 'Alternatives to the Use of Fetal Bovine Serum: Human Platelet Lysates as a Serum Substitute in Cell Culture Media', C. Rauch, E. Feifel, E. Amann 2, H. Spötl 2, H. Schennach 2, W. Pfaller and G. Gstraunthaler, ALTEX 28(4), 305–316, http://www.altex.ch/resources/altex_2011_4_305_316_Rauch1.pdf

p. 119 **FBS costs** Mark Post estimates it takes fifty litres.

Chapter Seven

pp. 136–7 **Decades of commercial fishing** 'The State of World Fisheries and Aquaculture: Meeting the Sustainable Development Goals', Food and Agriculture Organization of the United Nations, 2018, http://www.fao.org/3/i9540en/I9540EN.pdf

p. 137 **having to sail further out** D. Tickler, J. J. Meeuwig, M.-L. Palomares, D. Pauly and D. Zeller, 'Far from home: Distance patterns of global fishing fleets', Science Advances, 1 August 2018, http://advances.sciencemag.org/content/4/8/eaar3279

p. 137 **'bycatch'** R. W. D. Davies, S. J. Cripps, A. Nickson and G. Porter, 'Defining and estimating global marine fisheries bycatch', *Marine Policy*, Volume 33, Issue 4, July 2009, pp. 661–72, https://www.sciencedirect.com/science/article/pii/S0308597X09000050

p. 137 **We eat more fish** 'Global and regional food consumption patterns and trends: Availability and consumption of fish', World Health Organization, https://www.who.int/nutrition/topics/3_foodconsumption/en/index5.html

p. 142 **Mike once told a reporter** The *Sunday Times'* Danny Fortson, in the *Danny in the Valley* podcast: https://player.fm/series/danny-in-the-valley/finless-foods-mike-selden-we-brew-fish-meat

p. 147 **a 2010 paper** Dr Matthew Cole, 'Is in vitro meat the future of food? The case against', paper presented at the Vegetarian Society AGM, 11 September 2010, https://www.vegansociety.com/whats-new/news/vitro-meat-distraction-veganism

Chapter Eight

p. 157 **living mouse tissue** John Schwartz, 'Museum Kills Live Exhibit', *New York Times*, 13 May 2008, https://www.nytimes.com/2008/05/13/science/13coat.html

p. 160 **'I'm not concerned** Bruce Friedrich, 'Op-Ed: Is in vitro Meat the new in vitro fertilization?', *Los Angeles Times*, 25 July 2018, https://www.latimes.com/opinion/op-ed/la-oe-friedrich-ivmeat-20180725-story.html

p. 160 **clean meat produces more greenhouse gases** C. S. Mattick, A. E. Landis, B. R. Allenby and N. J. Genovese, 'Anticipatory Life Cycle Analysis of In Vitro Biomass Cultivation for Cultured Meat Production in the United States', *Environmental Science & Technology*, Volume 49, Issue 19, September 2015, https://pubs.acs.org/doi/ipdf/10.1021/acs.est.5b01614; H. L. Tuomisto and M. Joost Teixeira de Mattos,

'Environmental Impacts of Cultured Meat Production', *Environmental Science & Technology*, Volume 45, Issue 14, June 2011, https://pubs.acs.org/doi/abs/10.1021/es200130u; S. Smetana, A. Mathys, A. Knoch and V. Heinz, 'Meat alternatives: life cycle assessment of most known meat substitutes', *International Journal of Life Cycle Assessment*, Volume 20, September 2015, https://link.springer.com/article/10.1007%2Fs11367-015-0931-6

p. 160 **one study** P. Alexander, C. Brown, A. Arneth, C. Dias, J. Finnigan, D. Moran and M. D. A. Rounsevell, 'Could consumption of insects, cultured meat or imitation meat reduce global agricultural land use?', *Global Food Security*, Volume 15, December 2017, pp. 22–32, https://www.sciencedirect.com/science/article/pii/S2211912417300056

p. 162 **'challenges in cellular agriculture'** N. Stephens, L. Di Silvio, I. Dunsford, M. Ellis, A. Glencross and A. Sexton, 'Bringing cultured meat to market: Technical, socio-political, and regulatory challenges in cellular agriculture', *Trends in Food Science & Technology*, Volume 78, August 2018, pp. 155–66, https://www.sciencedirect.com/science/article/pii/S0924224417303400?via=ihub

Chapter Nine

p. 186 **A study** 'Pregnancy and maternity discrimination research findings', Equality and Human Rights Commission, https://www.equalityhumanrights.com/en/managing-pregnancy-and-maternity-workplace/pregnancy-and-maternity-discrimination-research-findings

p. 186 **In the US** 'By the Numbers: Women Continue to Face Pregnancy Discrimination in the Workplace', National

Partnership for Women & Families, October 2016,
http://www.nationalpartnership.org/our-work/resources/
workplace/pregnancy-discrimination/by-the-numbers-
women-continue-to-face-pregnancy-discrimination-in-the-
workplace.pdf

p. 188 **'she looked with contempt**   Genesis 16:2–4.

p. 189 **Philadelphia in 1884**   G. G. Mukherjee and B. N.
Chakravarty, *IUI: Intrauterine Insemination* (Jaypee Brothers
Medical Publishers, 2012), p. 383.

pp. 189–90 **in 2014**   Tamar Lewin, 'Coming to U.S. for Baby,
and Womb to Carry It', *New York Times*, 5 July 2014,
https://www.nytimes.com/2014/07/06/us/foreign-
couples-heading-to-america-for-surrogate-pregnancies.
html

p. 190 **in 2018**   Valeria Perasso, 'Surrogate mothers: "I gave
birth but it's not my baby"', BBC News, 4 December 2018,
https://www.bbc.co.uk/news/world-46430250

p. 190 **There are far too many**   Matthew Renda, 'Surrogate
Mother's Attempt to Regain Her Children Fails in Ninth
Circuit', Courthouse News Service, 12 January 2018, https://
www.courthousenews.com/surrogate-mothers-attempt-to-
regain-her-children-fails-in-ninth-circuit/; 'Luca's Law' blog,
https://lucaslaw.blog/

p. 191 **The judge said**   'Baby Gammy: Surrogacy row family
cleared of abandoning child with Down syndrome in
Thailand', ABC News, 14 April 2016, https://www.abc.net.au/
news/2016-04-14/baby-gammy-twin-must-remain-with-
family-wa-court-rules/7326196

pp. 191–2 **Now Ukraine**   Kevin Ponniah, 'In search of
surrogates, foreign couples descend on Ukraine', BBC News,
13 February 2018, https://www.bbc.co.uk/news/world-
europe-42845602

p. 192 **In December 2015** 'Parliamentary questions: Question for written answer P-005909/2016/rev.1 to the Commission', European Parliament, 18 July 2016, http://www.europarl. europa.eu/doceo/document/P-8-2016-005909_ EN.html?redirect

Chapter Ten

p. 195 **in April 2017** E. A. Partridge, M. G. Davey, M. A. Hornick, P. E. McGovern, A. Y. Mejaddam, J. D. Vrecenak, C. Mesas-Burgos, A. Olive, R. C. Caskey, T. R. Weiland, J. Han, A. J. Schupper, J. T. Connelly, K. C. Dysart, J. Rychik, H. L. Hedrick, W. H. Peranteau and A. W. Flake, 'An extra-uterine system to physiologically support the extreme premature lamb', *Nature Communications*, Issue 8, 25 April 2017, https://www.nature.com/articles/ncomms15112

p. 199 **But 87 per cent** Gene Emery, 'Survival rates for extremely preterm babies improving in U.S.', Reuters, 15 February 2017, https://www.reuters.com/article/us-health-preemies-survival-impairments/survival-rates-for-extremely-preterm-babies-improving-in-u-s-idUSKBN15U2SA

p. 199 **chronic lung disease** B. J. Stoll, N. I. Hansen, E. F. Bell et al., 'Trends in Care Practices, Morbidity, and Mortality of Extremely Preterm Neonates, 1993–2012', *JAMA*, Volume 314, Issue 10, 8 September 2015, https://jamanetwork.com/journals/jama/fullarticle/2434683

p. 199 **between 1995 and 2006** T. Moore, E. M. Hennessy, J. Myles, S. J. Johnson, E. S. Draper, K. L. Costeloe and N. Marlow, 'Neurological and developmental outcome in extremely preterm children born in England in 1995 and 2006: the EPICure studies', *BMJ*, Issue 345, 4 December 2012, https://www.bmj.com/content/345/bmj.e7961

p. 199 **the number of children** *Born Too Soon: The Global Action Report on Preterm Birth*, March of Dimes, The Partnership for Maternal, Newborn & Child Health, Save the Children, World Health Organization (WHO Publications, 2012), https://www.marchofdimes.org/ materials/born-too-soon-the-global-action-report-on-preterm-.pdf; K. L. Costeloe, E. M. Hennessy, S. Haider, F. Stacey, N. Marlow and E. S. Draper, 'Short term outcomes after extreme preterm birth in England: comparison of two birth cohorts in 1995 and 2006 (the EPICure studies)', *BMJ*, Issue 345, 4 December 2012, https://www.bmj.com/ content/345/bmj.e7976

p. 199 **Preterm birth** 'Facts about EVE Therapy and extreme preterm birth: FAQ about EVE Therapy – The Artificial Womb', Women & Infants Research Foundation, Western Australia, http://www.tohoku.ac.jp/en/press/images/ artificial_womb_faq.pdf

p. 205 **Haldane imagined an essay** J. B. S. Haldane, 'Daedalus, or Science and the Future', 4 February 1923, http://bactra.org/ Daedalus.html

pp. 206–7 **'We have now reached a stage** *Manifesto*, Gay Liberation Front, London, 1971, https://sourcebooks. fordham.edu/pwh/glf-london.asp

p. 214 **in 2016** M. N. Shahbazi, A. Jedrusik, S. Vuoristo et al., 'Self-organization of the human embryo in the absence of maternal tissues', *Nature Cell Biology*, Issue 18, 4 May 2016, https://www.nature.com/articles/ncb3347

p. 215 **The fourteen-day deadline** I. Hyun, A. Wilkerson and J. Johnston, 'Embryology policy: Revisit the 14-day rule', *Nature*, Volume 533, Issue 7602, 4 May 2016, https://www. nature.com/news/embryology-policy-revisit-the-14-day-rule-1.19838-/agreement

p. 215 **on an extrauterine scaffold** 'Ability of three-dimensional (3D) engineered endometrial tissue to support mouse gastrulation in vitro', Liu, Hung-Ching et al., *Fertility and Sterility*, Volume 80, 78, https://www.fertstert.org/article/S0015-0282(03)02008-9/fulltext

Chapter Eleven

p. 217 **Immaculate gestation** This term was coined by Scott Gelfand and John Shook in their book *Ectogenesis: Artificial Womb Technology and the Future of Human Reproduction* (Rodopi, 2006).

p. 217 **artificial wombs** 'The Moral Imperative for Ectogenesis', *Cambridge Quarterly of Healthcare Ethics* (2007), 16, 336–345; and 'In Defence of Ectogenesis', *Cambridge Quarterly of Healthcare Ethics* 21 (2012): 90–103.

p. 224 **in 2013** Sophie Borland, 'Doctors and nurses "don't need to show compassion": Academic says staff should be able to carry out daily tasks without being kind to patients', *Daily Mail*, 18 September 2013, https://www.dailymail.co.uk/news/article-2424063/Academic-claims-doctors-nurses-dont-need-compassion-patients.html

p. 225 **Since 2001** James Gallagher, 'First baby born after deceased womb transplant', BBC News, 5 December 2018, https://www.bbc.co.uk/news/health-46438396

p. 229 **Japanese scientists** Philip Ball, 'Reproduction revolution: how our skin cells might be turned into sperm and eggs', *Guardian*, 14 October 2018, https://www.theguardian.com/science/2018/oct/14/scientists-create-sperm-eggs-using-skin-cells-fertility-ethical-questions

p. 233 **in 2016** Juno Roche, 'My Longing To Be A Mother, As A Trans Woman', Refinery29, 8 September 2016, https://www.refinery29.com/en-gb/trans-woman-motherhood

Chapter Twelve

p. 241 **Barbara has bought** 'Statistics', Project Prevention, http://projectprevention.org/statistics/

p. 243 **By 2015** 'Special report: Alabama leads nation in turning pregnant women in to felons', AL.com, 23 September 2015, https://www.al.com/news/2015/09/when_the_womb_is_a_crime_scene.html

p. 243 **Prenatal cocaine exposure** J. P. Ackerman, T. Riggins and M. M. Black, 'A Review of the Effects of Prenatal Cocaine Exposure Among School-Aged Children', *Pediatrics*, Volume 125, Issue 3, March 2010, pp. 554–65, https://www.ncbi.nlm.nih.gov/pmc/articles/PMC3150504/

p. 245 **When the few details** Colin Freeman, 'Child taken from womb by caesarean then put into care', *Telegraph*, 30 November 2013, https://www.telegraph.co.uk/news/uknews/law-and-order/10486452/Child-taken-from-womb-by-caesarean-then-put-into-care.html and https://www.telegraph.co.uk/comment/columnists/christopherbooker/10485281/Baby-forcibly-removed-by-caesarean-and-taken-into-care.html

p. 245 **Between 2008 and 2014** I. Jensen, A. Fredrikstad, S. Saabye and P. Haugen, 'Child welfare takes three times as many newborns', TV 2 News, 13 April 2016, https://translate.google.com/translate?hl=en&sl=auto&tl=en&u=https%3A%2F%2Fwww.tv2.no%2Fnyheter%2F8219203%2F

p. 245 **By far the most common reason** I. P. Nuse, 'Protests mount against Norwegian Child Welfare Service', ScienceNordic, 10 February 2018, http://sciencenordic.com/protests-mount-against-norwegian-child-welfare-service

p. 245 **'lack of parenting skills'** Tim Whewell, 'Norway's Barnevernet: They took our four children . . . then the baby', BBC News, https://www.bbc.co.uk/news/magazine-36026458

Chapter Thirteen

p. 269 **In 2015** 'Largest Ever Poll on Assisted Dying Finds Increase in Support to 84% of Britons', Dignity in Dying press release, 2 April 2019, https://www.dignityindying.org.uk/news/ poll-assisted-dying-support-84-britons/

p. 269 **4 per cent of all deaths** 'Dutch Regional Euthanasia Review Committee Annual Report 2018', https://english. euthanasiecommissie.nl/the-committees/documents/ publications/annual-reports/2002/annual-reports/ annual-reports

p. 271 **In 1945** These figures come from US national statistics quoted in Atul Gawande's *Being Mortal* (Macmillan USA, 2014), essential reading for anyone seeking to understand how technology has changed the meaning of death.

p. 271 **the option of killing themselves** Philip Nitschke, 'Euthanasia is a rational option for prisoners facing the torture of life in jail', *Guardian*, 27 September 2014, https://www.theguardian.com/ commentisfree/2014/sep/27/euthanasia-is-a-rational-option- for-prisoners-facing-the-torture-of-life-in-jail

Chapter Fourteen

p. 285 **The patient** Lisa Belkin, 'Doctor Tells of First Death Using His Suicide Device', *New York Times*, 6 June 1990, https://www.nytimes.com/1990/06/06/us/doctor-tells-of- first-death-using-his-suicide-device.html

p. 286 **Most of his patients** L. A. Roscoe, J. E. Malphurs, L. J. Dragovic and D. Cohen, 'A Comparison of Characteristics of Kevorkian Euthanasia Cases and Physician- Assisted Suicides in Oregon', *Gerontologist*, Volume 41, Issue 4, 1 August 2001, pp. 439–46, https://academic.oup.com/ gerontologist/article/41/4/439/600708

p. 286 **autopsies showed**　A lot of this comes from work done by the Detroit Free Press, referenced here: *Update*, Volume 25, Issue 3, Patients Rights Council, 2011, http://www. patientsrightscouncil.org/site/wp-content/uploads/2011/07/ Update_2011_3.pdf

p. 289 **'It's not rocket science**　'Nitschke launches $50 death machine', *Sydney Morning Herald*, 18 November 2003, https://www.smh.com.au/national/nitschke-launches-50-death-machine-20031118-gdhss2.html

p. 293 **'Witlessly infantile**　Mark Monahan, 'Edinburgh 2015: Dicing With Dr Death, The Caves, review: "witlessly infantile"', *Daily Telegraph*, 8 August 2015, https://www.telegraph.co.uk/ theatre/what-to-see/edinburgh-2015-dr-death/

p. 293 **'Laughs were sparing**　Cameron Woodhead, 'Melbourne International Comedy Festival review: No one dying of laughter in Philip Nitschke's Dicing With Death', *Sydney Morning Herald*, 4 April 2016, https://www.smh.com.au/ entertainment/comedy/melbourne-international-comedy-festival-review-no-one-dying-of-laughter-in-philip-nitschkes-dicing-with-death-20160404-gny6oz.html

p. 297 **'Meet the Elon Musk**　Nicole Goodkind, 'Meet the Elon Musk of Assisted Suicide, Whose Machine Lets You Kill Yourself Anywhere', *Newsweek*, 1 December 2017, https:// www.newsweek.com/elon-musk-assisted-suicide-machine-727874

Chapter Fifteen

p. 308 **Huffington Post**　Philip Nitschke, 'Here's Why I Invented A "Death Machine" That Lets People Take Their Own Lives', HuffPost, 4 May 2018, https://www.huffpost.com/ entry/sarco-death-philip-nitschke_n_5abbb574e4b03e2a5c 7853ca

p. 308 *Vice*   Matt Shea, ' "Dr Death" Has a New Machine That's Meant to Disrupt the Way We Die', *Vice*, 10 May 2019, https://www.vice.com/en_uk/article/5979qd/sarco-euthanasia-machine-philip-nitschke

p. 310 **'Welcome to the Monkey House'**   Kurt Vonnegut, *Welcome to the Monkey House* (Delacorte Press, 1968).

p. 310 **venture capitalists funding anti-ageing research**   The Longevity Fund is leading the way with this: https://www.longevity.vc/

p. 313 *Gelderlander*   Paul Bolwerk, 'Noa (16) uit Arnhem is nu al klaar met haar verwoeste leven', *Gelderlander*, 1 December 2018, https://www.gelderlander.nl/home/noa-16-uit-arnhem-is-nu-al-klaar-met-haar-verwoeste-leven~a01a7bd1/

p. 313 **Philip later wrote**   'The Death of Noa Pothoven', Peaceful Pill Handbook blog, 5 June 2019, https://www.peacefulpillhandbook.com/the-death-of-noa-pothoven/

p. 326 **a 12 per cent rise**   S. Stack, 'Media coverage as a risk factor in suicide', *Journal of Epidemiology & Community Health*, Issue 57, 1 April 2003, pp. 238–40, https://jech.bmj.com/content/57/4/238.full

p. 326 **a 10 per cent rise**   D. S. Fink, J. Santaella-Tenorio, K. M. Keyes, 'Increase in suicides the months after the death of Robin Williams in the US', *PLOS ONE*, Volume 13, Issue 2, 7 February 2018, https://journals.plos.org/plosone/article?id=10.1371/journal.pone.0191405

Epilogue

p. 331 **In October 2019**   Bethany Muller, 'Artificial womb to be developed for premature babies', BioNews, 14 October 2019, https://www.bionews.org.uk/page_145518

p. 331 **in September 2019** Bruce Friedrich, 'Cultivated Meat: Why GFI Is Embracing New Language', Good Food Institute, 13 September 2019, https://www.gfi.org/cultivatedmeat

p. 331 **Beyond Meat shares** 'Beyond Meat shares extend gains to over 600% since IPO', *Financial Times*, https://www.ft.com/content/df314088-8b91-11e9-a24d-b42f641eca37

p. 332 **Most of those who died** L. A. Roscoe, J. E. Malphurs, L. J. Dragovic and D. Cohen, 'A Comparison of Characteristics of Kevorkian Euthanasia Cases and Physician-Assisted Suicides in Oregon', *Gerontologist*, Volume 41, Issue 4, 1 August 2001, https://academic.oup.com/gerontologist/article/41/4/439/600708

p. 333 **women choose it more often** Rachael Wong, 'We need to address questions of gender in assisted dying', The Conversation, 24 October 2017, http://theconversation.com/we-need-to-address-questions-of-gender-in-assisted-dying-85892

p. 333 **men eat more meat** Hamish J. Love and Danielle Sulikowski, 'Of Meat and Men: Sex Differences in Implicit and Explicit Attitudes Toward Meat', *Frontiers in Psychology*, Volume 9, 20 April 2018, https://www.ncbi.nlm.nih.gov/pmc/articles/PMC5920154/